A 2-HOUR MOVIE SCRIPT

CYBERDROID:

THE FUTURE OF THE WORLD

By: Tina Schaneville

A 2-HOUR MOVIE SCRIPT

CYBERDROID:

THE FUTURE OF THE WORLD

By: Tina Schaneville

Tina Schaneville

P.O. Box 482

Chalmette, La. 70044

1-706-360-3410

Cyber4us1@gmail.com

Library of Congress ISBN # 9780615626840

Dedication

I dedicate this book to the Creator of the Universe, my children, family, and friends. May the Creator of the Universe bless us all! :)

04-01-2012

Tina currently resides in Louisiana, United States of America with her daughter and two Chihuahuas. She is currently attending an accredited university to obtain her Ph.D. degree in the area of Information Technology. Tina is near the end of her Ph.D. program. She conducts research in her spare time to try to make the world a better place in which to live.

Tina Schaneville

P.O. Box 482

Chalmette, La.   70044

1-706-360-3410

Cyber4us1@gmail.com

Table of Contents

Setting: Cyberdroid Corporation, a large corporation, which makes desktop computers, laptops, software, and androids.

Time: Year 2012

Where: Location in the United States of America

Classification: This 2 - hour movie script contains a love story, murder mystery, and conflict between androids and human beings.

2 - Hour movie script rating: PG17

Cast of Characters @ Cyberdroid Corporation:

Chief Executive Officer of Cyberdroid Corporation - Blake Fencington, Ph.D.

Blake is a Caucasian man who is 6'0" ft. tall with light brown thick straight hair combed to the side, clear skin, blue eyes, a small straight nose, and a masculine chiseled chin. His voice is a masculine, deep smooth voice. He has a Ph.D. in the area of Information Technology. He has a tall thin body type.

Chief Information Technology Officer - Ms. Jennifer Smith, Ph.D.

Jennifer is a Caucasian lady who is 5'02" tall with a light clear tan complexion, a small nose, and full feminine lips. She has thick beige blond, long wavy hair with a side part. Her eyes are brown. She takes good care of herself by counting calories, doing aerobic exercise, and taking a daily multivitamin. She has medium sized breasts and is in excellent shape.

Co-Chief Information Technology Officer - Colby James, Ph.D.

Colby is a Caucasian man who is 5'10" ft. tall with dark blond straight hair combed to the side, fair clear skin, blue eyes, a small straight nose, and a masculine chin. His voice is deep, smooth, and masculine. He has a Ph.D. in the area of Information Technology. He has a tall toned body type.

Chief Financial Officer - Joel Milson, Ph.D.

Joel is a Caucasian man 5'11" ft. tall with blue/green eyes, light brown hair, clear light toned skin, small nose, with a short hairstyle combed to the side. He has a masculine voice. He has a Ph.D. degree in Information Technology and Business Management. He has a tall thin body type.

Cyberdroid's Secretary to the CEO, Ken – Ms. Jollyday

Ms. Nicole Jollyday is a 60-year-old lady with all white hair that is worn in a French twist on the back of her head. She is 5'0" ft. tall with blue eyes. She has a very thin body structure.

A client of Cyberdroid Corporation – Mr. Yoko Yenni

Mr. Yoko Yenni is a Chinese man with white straight thinning hair. He has pleasant facial features. He is a regular purchaser of androids for China. He is 5'05" ft. tall with brown eyes. He has a thin body type. He is 65 yrs. old.

Mr. Yoko Yenni's lawyer – Mr. Vay Vo

Mr. Vay Vo is of Chinese/French decent. He has almond shaped eyes, a small nose and green eyes. He is 5'10" tall. He has thick dark brown straight hair combed to the side. He is Mr. Yoko Yenni's corporate lawyer.

Dr. Charie Ballof is a Caucasian lady who is 5'02" ft. tall with a slender body frame. She has thick long straight light brown hair to her but. She has a small nose and light brown eyes. She has full luscious lips. She is beautiful. She has a Ph.D. in Information Technology research.

Former President Strong-tree

President Strong-tree is a Caucasian man with blue eyes and light brown straight hair that is combed to the side. He is 50 years old. He has an average body build. He resides in Texas with his wife, Barelle.

Barelle- Former President Strong-tree's wife

Barelle has short light brown hair and blue eyes. She has a thin body frame.

Scientist Glenn Ganuchi is 5'09 ft. tall. He has white hair that is unruly and all one length to his neck. He has an average body build. He has hazel eyes. He resides in Beijing, China. He is lead scientist for the Cyberdroid Corporation.

Judge Jack Murphy

Judge Jack Murphy is the judge in Division C at the Atlanta, Georgia, United States Courthouse. He presides over civil matters mostly. Corporate law is his specialty. He has light brown, straight hair combed to the side, and light brown eyes. He is 58 years old. Judge Jack is 6'0" ft. tall with an average body frame.

Bailiff Joe Jackson

Bailiff Joe Jackson is a tall, broad shouldered man. He is 6'03" ft. tall with bulging muscles. He lifts weights regularly to body build. He has brown eyes and wavy grey/brown thinning hair combed straight back. Bailiff Joe is 60 years old. Bailiff Joe wears a gun and a holster. He keeps order in the courtroom. He detains people who are rowdy in the courtroom until the police come to get the man or lady causing trouble.

Cyberdroid's Lawyer – Christopher Murphy

Lawyer Christopher Murphy is the lawyer for Cyberdroid Corporation. He handles all legal matters for the Cyberdroid Corporation. Christopher has bright green eyes and medium brown, thick straight hair combed to the side. He is 6'0" ft. tall. He has excellent male facial features such as a masculine chin. Lawyer Christopher is 46 years old.

Inspector Cluenoo: He is 5'07" tall. He has black straight hair combed to the side. He is 58 years old. He has a long work history as an inspector. He does his job correctly. However, he is clumsy. Inspector Cluenoo chews gum constantly. He smokes cigars. He has a pot beer belly.

Dr. Williams: Dr. Williams is a medical general practitioner doctor. He is 5'10" ft. tall with blue/green eyes and light brown hair combed to the side. Dr. Williams has a medium, average body build. He is Blake Fencington's college friend. He is 46 yrs. old.

Cast of Characters @ Collins Corporation

Chief Executive Officer (CEO) of Collins Corporation – Ken Collins, Ph.D.

Ken Collins, Ph.D. is 6'01" tall. He has a slender body frame. He has light brown hair with blue eyes. He has nice facial features. Ken is 46 years old.

Chief of Information Technology – Chrissie Belle, Ph.D.

Chrissie Belle, Ph.D. has light brown, long, straight thick hair combed to the side with a small straight nose, straight white teeth, clear skin, and full luscious lips. She is petite and is 5'02" ft. tall. She is 26 years old.

Chief Financial Officer – Michael Bennett, Ph.D.

Michael Bennett, Ph.D. is 6'0" ft. tall with a medium build. He has blond beige hair combed to the side with light green eyes. He is 57 years old.

 Private Investigator - Barry Benad

Barry Benad is 6 ft. tall with blue eyes and an average build body type. He has all white short hair that is combed straight back. He has earned a Bachelor Degree. Barry has taken many classes at the Federal Bureau of Investigation Academy. He is 50 years old.

Collins' Corporation Office Worker - Kelly Crespo

Collins' Corporation Office Worker - Kelly Crespo is 5'08" ft. tall with blue eyes and blond hair. She has a slender body type. She has earned a Bachelor degree in business administration. Kelly is 28 years old.

Collins' Corporation Office Worker - Amanda Wilson

Collins' Corporation Office Worker - Amanda Wilson is 5'07" ft. tall with dark brown hair and brown eyes. She has a large frame body type. She has earned a Bachelor degree in business administration. Amanda is 32 years old.

Manager in Business Administration Department at Collins' Corporation - Simon Simmons

Simon Simmons is 5'07" ft. tall with thin grey hair combed over to the side. He has a large belly, and a wide body frame. He has hazel eyes, and speaks English with a French Accent. He is 55 years old.

Racio Guerra – Mr. Barry Benad's clumsy, clueless, and foreign accomplice

Racio Guerra is 5'04" ft. tall with a slender body. He has black, curly hair a little passed his neck. He has a flat, wide nose with regular lips. His eye color is dark brown. His skin tone is a dark tan color. He is of Spanish and Mexican decent. He has a Mexican accent and speaks broken English. He is Barry Benad's accomplice. He is 27 years old.

Collins Corporation Lawyer – Bill Walker

Collins Corporation Lawyer – Bill Walker is 5'08" ft. tall with a heavyset body frame. He has blue eyes with short, white hair combed straight back. He smokes cigars when not in the courtroom. He wears a light tan cowboy hat and cowboy boots with his coat and tie. He takes his cowboy hat off while in the courtroom. Lawyer Bill is 62 years old.

Policeman Henry Sullivan – Policeman Henry Sullivan is 5'09" ft. tall with a large body frame. He has thin, straight grey hair combed over to the side. He has blue eyes, a short nose, and chubby cheeks.

The Cyberdroid Corporation manufactures desktop computers, laptops, software, and androids. It was incorporated as a corporation in 1987. Cyberdroid has grown into a worldwide conglomerate. It supplies many stores with computer devices and androids.

Androids have become so human looking that no one can tell them apart from human beings. Their android identity is kept secret because human beings are in conflict with them. Humans do not like them because they are competing and applying for human job positions. Androids have been given all the rights of human beings in the year 2012 from the Supreme Court of the United States of America because they can now fully think for themselves just like humans. The only difference is that they have machinery inside for body parts along with real blood. They have a DNA code and a fully self-governing brain. The justices of the United States Supreme Court have given androids full human status because of the self-governing brain that works without any programming maintenance. Once the android brain has been created for the android, it evolves with DNA into a real functioning human brain. Therefore, that is why the androids have been given full human status and all rights of human beings in the year 2012.

The leading competitor of Cyberdroid Corporation is Collins Corporation. Collins Corporation only manufactures desktop computers and laptops. Collins Corporation is second only to Cyberdroid. It was incorporated into a corporation in 1990. Collins Corporation has a wonderful laptop computer with all the highest amount of graphics. It has an excellent market selling to computer users that love to play video games on the best graphical laptop and desktop computers.

1.) INSIDE CYBERDROID CEO OFFICE

The CEO's office is decorated in a modern style with light blue, light grey and silver. It is clean and organized. It is on the 20th floor of a high-rise office building. The office building is covered with glass and appears silver in the sunlight. It is a tall rectangular office skyscraper located in Atlanta, Georgia.

CEO Blake: (on phone) Yes, I know I need to ship 10,000 androids to China. Get the financial papers together for the meeting with Yoko Yenni. We will finalize the deal on Tuesday.

Joel: Yes, sir. I am on it for you. I will have all the papers ready to sign after the meeting with Mr. Cooper, the lawyer.

CEO Blake: Thanks, Joel. Keep up the excellent work! :)

2.) INSIDE CYBERDROID CEO OFFICE

Jennifer knocks on the glass office door.

CEO Blake: Please come in Jennifer. It is always a pleasure to see you.

Jennifer: Mr. Blake, I just want to show you some ideas I have come up with for the Information Technology Department. When you have some free time, please look over these papers and let me know in an email or meeting what you think.

CEO Blake: Yes, Jennifer, I most certainly will look these over and give you an honest opinion.

Jennifer: Thank you, Mr. Blake.

CEO Blake: Jennifer, please call me, 'Blake'.

Jennifer: Thanks, Blake. Have a nice day! :)

3.) Tuesday morning: Mr. Yoko arrives to sign the papers and finalize the deal.

CEO Blake: Ah, Mr. Yoko, so nice to see you again.

Mr. Yoko Yenni: (Mr. Yoko has a Chinese accent.) It is a pleasure to see you, Mr. Fencington.

CEO Blake: Here are all of the papers. Please read them and sign on all the dotted lines. Please call me, Blake.

Mr. Yoko Yenni: Ok. Yes, but before I sign. I must call my assistant. He is a lawyer. He will advise me.

CEO Blake: Yes, by all means, do call him to come in here to assist you.

Mr. Yoko Yenni: Thanks, Blake.

A Chinese man with black straight hair enters the room.

Mr. Yoko Yenni: Vo, please look these papers over and tell me what you think.

Vay Vo: Yes, Mr. Yoko.

After 30 minutes, Mr. Vo has an answer for Mr. Yoko.

Vay Vo: Mr. Yoko, it appears to be all in order. You can sign.

Mr. Yoko Yenni: Thanks, Vo. You are free to go now. I will call you later.

Vay Vo: Thanks, Mr. Yoko.

CEO Blake: I am glad we are all on the same page, so to speak. Mr. Yoko, please sign on all of the dotted lines.

Mr. Yoko Yenni: Yes, Blake it is always a pleasure doing business with you!

CEO Blake: I will ship the 10, 000 androids by ship on Thursday. I will have high security on board. No need to worry, Mr. Yoko. All will be fine.

Mr. Yoko Yenni: Thanks, Blake. I appreciate it. I will keep in touch.

CEO Blake: Thanks to you, Mr. Yoko. I will keep in touch as well.

4.) Monday morning in CEO Blake's office

CEO Blake: (on phone) Jennifer, please come into my office.

Jennifer: (on phone) Yes, Blake.

CEO Blake: These are wonderful ideas! I could not have done a better job myself. Please implement these ideas as soon as possible.

Jennifer: I am so glad that you like these ideas, Blake.

CEO Blake: Jennifer, I would like to take you out to dinner and a movie on Saturday night. Are you available?

Jennifer: Yes, Blake. I would be delighted! What time will you pick me up?

CEO Blake: Well, 4 p.m. Is that fine? That way we can go to the 7 p.m. movie.

Jennifer: Yes, Blake that is fine.

CEO Blake: See you then. Put on your best attire. I will make reservations for us at Anjou's Restaurant.

Jennifer: Yes, thanks. I will be ready. (smiling)

5.) Saturday at 4 p.m. in front of Jennifer's house at 207 Magnolia Tree Lane. It is a tan brick house with two large fully-grown Magnolia trees on the front lawn.

CEO Blake: Blake walks up the walkway to Jennifer's house. He knocks on the door.

Jennifer: Jennifer opens the door looking quite beautiful in a dark blue velvet dress that fits her skin tight around the breasts and waist. Then, the dress fans out and stops at knee length. Jennifer looks so beautiful!

CEO Blake: Jennifer, you look gorgeous!

Jennifer: Thanks, Blake. Just let me put the burglar alarm on.

CEO Blake: Ok, I will be waiting in the Beep.

Jennifer checks the house to make certain that all lights are off; the house is locked, and sets the burglar alarm. Then, she walks out to Blake's light blue/silver Beep Grand SUV. Jennifer gets in the passenger front seat next to Blake.

CEO Blake: Jennifer, we are going to have a wonderful time!

Jennifer: Yes, I am certain we will. (smiling)

CEO Blake drives carefully to Anjou's Restaurant. The valet parks the light blue/silver Beep Grand SUV. Blake puts his arm around Jennifer's bare shoulders and walks with her to Anjou's Restaurant, which is only 10 feet away. The weather was just perfect! It was April 1st, 2012, with clear spring-like weather in the 80's. Jennifer was ecstatic to be on a date with Blake.

The waiter shows the couple to a table in the corner of the restaurant. It smells wonderful of different kinds of foods. The restaurant is immaculate with a nice décor in different pastel colors. As always, the food is delicious! Jennifer ordered the grilled chicken salad with light Caesar dressing. Blake ordered grilled salmon, baked potato, and a side salad with French dressing. Blake ordered Iced tea with lemon and aspartame sweetener. Jennifer ordered diet root beer with a little ice.

The dinner went well. The conversation was friendly. It was a pleasant evening. Blake drove Jennifer home. He kissed her on the cheek, and put his arm around her shoulder. Both Blake and Jennifer felt happy, adored, and special!

Blake: Goodnight, Jennifer. See you on Monday at the office.

Jennifer: Goodnight, Mon Ami. Au revoir (French for until we meet again).

6.) Meanwhile, at Collins Corporation (Cyberdroid's leading competitor):

Ken: I wonder what Blake is doing over at Cyberdroid Corporation. He has been pestering me like a fly for my entire business career. I try to get ahead of him, but I am always second. If it is the last thing that I ever do, I will defeat him!

Chrissie:
(Chrissie put her arm around Ken's neck.) Do not let it get the best of you, Ken. I am certain that one day you will succeed over him and the Cyberdroid Corporation. Just calm down. It is not worth it. You will get an ulcer or a heart attack over this.

Ken:
Yes, Chrissie, you are always correct. I cannot let this affect my health. I have responsibilities.

Chrissie:
That is better. Just relax and sit down. There is a silver lining and rainbow coming your way. Just have a positive attitude and personality.

Chrissie leaves the CEO room at Collins Corporation.

7.) CEO Ken at Collins Corporation on phone:

This is Ken at Collins Corporation. How is your family doing?

Barry: Well, I just got divorced recently. I am living with my youngest son only now. He is 17 years old. The other three sons are grown up now.

CEO Ken: Oh, I am sorry to hear that.

Barry: It is fine. We have adjusted to the situation.

CEO Ken:

Barry, what are you doing now?

Barry: I am working on a case about finding a 300 LB. lady who is missing. Her name is Marie Miller. Her maiden name is Marie Casort. She is 56 years old with brown/ grey hair and blue eyes. Her grown brothers miss her terribly!

CEO Ken: Oh, that is a shame. I hope you find her. I am just wondering how a 300 LB. lady can end up missing. It would be difficult to hide. LOL… Ha ha.

Barry: Good one, Ken! LOL… Ha ha…

What can I do for you, Ken?

CEO Ken: I need you to spy for me at the Cyberdroid Corporation. More specifically, I need to know how the androids were created or a blueprint. I saw Blake on Happy Morning United States. He told the newscaster that he had a divine experience. Can you believe that?

Barry: Well, I will have to think about that. Yes, Ken, I will work on that for you.

CEO Ken: Thanks, Barry, I can always count on you! Call me when you find something or regularly just to check in with me.

Barry: Yes, I will do that. I will do my best.

CEO Ken: Ok, you have a good evening. Tell your son that I said, 'hello'. Bye, Barry.

Barry: Bye, Ken.

8.) Monday morning at Cyberdroid Corporation

CEO Blake: (on phone) Jennifer, how are you?

Jennifer: Just fine. I really enjoyed our date.

CEO Blake: I would like to see you on a regular basis.

Jennifer: I would love to date you! I am presently available.

CEO Blake: Fantastic! You made my day! Is Saturday fine for our dates?

Jennifer: Yes, I will see only you.

CEO Blake: I truly like you, Jennifer.

Jennifer: Thanks, Blake.

CEO Blake: I would like to take you to lunch sometimes.

Jennifer: That is fine, also.

Blake: Have a nice day, Jennifer.

Jennifer: You too, Blake.

9.) Collins Corporation in the Business Administration Department

Kelly Crespo: I really hate androids! They are taking all of the promotions, jobs, and opportunities in life.

Amanda Wilson: Yes, and a person cannot tell if they are an android unless they find the android key that plugs into the wall for some electricity. The electricity is what keeps them going. I wonder sometimes if I would really like to live forever, and keep plugging into the wall for electricity regularly.

Kelly Crespo: Well, I do not like the androids' longevity life span. We human beings can only live 100 years or a little older. That is if we take good care of ourselves.

Amanda Wilson: I know. I do not like the androids' longevity either. They are taking away jobs and opportunities from honest, diligent human beings.

Kelley Crespo: There is nothing we can do about it. A Supreme Court of the United States' judge granted the androids the same inalienable rights as human beings.

Amanda Wilson: The judge must have been senile or something.

Kelly Crespo: Shh! Here comes the manager, Mr. Simon Simmons.

Collins' Corporation Manager: Simon Simmons

Well, ladies, how is it going today?

Kelly Crespo: Fine. Yes, we are terrific, diligent workers, Mr. Simmons.

Simon: Keep up the terrific work, ladies!

10.) At Cyberdroid Corporation

Blake walks to Jennifer's office inside the Cyberdroid Corporation.

CEO Blake knocks on the door. (Knock. Knock.)

Jennifer: Who is it?

CEO Blake: It is I, Blake.

11.)   Jennifer: You may enter, Blake.

CEO Blake: Jennifer, how would you like to take a trip with me to Joyney World in Florida, United States of America?

Jennifer: Oh, I would love that. I have not been back to Joyney World since I was a child.

CEO Blake: Well, Joyney World has added on so much more since you were a child.

There are places inside Joyney World for horseback riding, bike riding, swimming in a clean manmade lagoon with man-made waterfalls. There are cottages for couples, and so much more!

Joyney World is for children and people of all ages. There is something for everyone at Joyney World.

Jennifer: Yes, I heard about that once. I viewed photos on the Internet.

CEO Blake: Let us go for a long weekend consisting of Thursday, Friday, Saturday, Sunday, and Monday.

Jennifer: Oh, that sounds, perfect!

CEO Blake: Blake puts his arm around Jennifer's neck and French kisses her.

Jennifer: Jennifer French kisses him back.

CEO Blake: I will fly us there in my SUV/plane. It is a rugged, all-terrain vehicle with wings that open up to turn into a plane.

Jennifer: Yes, I know. I heard about it on the news. Only wealthy people can afford to own one.

CEO Blake: I will pick you up at 10 a.m. on Thursday morning.

Jennifer: That is fine. I will go home to pack tonight.

CEO Blake: Marvelous! See you then. I have to go now. I have plenty of responsibilities at Cyberdroid Corporation.

11.) At Collins' Corporation:

Barry (thinking about Cyberdroid corporation)

How am I going to get through security at Cyberdroid Corporation?

12,) At Jennifer's House:

Blake picks Jennifer up promptly at 10 a.m. on Thursday morning. Blake told Jennifer to buckle her seatbelt. She acquiesced. They departed in the rugged, light silver/ blue, sleek vehicle/plane with aerodynamic styling.

CEO Blake: Here put these in your ears, if you need them. These are excellent for ear popping at high altitudes.

Jennifer: Ok, thanks, Blake.

At Joyney World:

13.)   They arrived at Joyney World after a 2-hour flight from Atlanta, Georgia, United States of America. Blake parked the SUV/plane on the special landing that is specially designated for private charter planes and SUV/planes. He got the luggage out of the back. Pressed the wing enter button. The plane wings receded into the SUV/plane. Blake locked the vehicle/plane and disembarked from it. He held out a hand for Jennifer so she would not fall when she stepped down from the vehicle/plane. There are many types and models of vehicle/planes. However, Blake prefers the rugged terrain Beep SUV/plane for durability, space, dependability, and style.

14.)   A hotel worker greeted Blake and Jennifer. He said, "Follow me this way to your Hotel, which is located on the Joyney World property. I hope your stay will be a pleasant and enjoyable one". They walked about 100 steps. Then, boarded a monorail bus, which took them to the 'Happy Day' Hotel. It was about 12' o' clock lunchtime. However, both Blake and Jennifer were not hungry. Jennifer felt stressful because she is afraid of flying on planes. She just wanted to go take a nap and relax for a little while. Jennifer felt nauseous, also.

Blake and Jennifer stayed at the Contemporary Hotel. It had the monorail there. Blake let Jennifer rest in her adjoining hotel suite next to his. He called her 1 hour later to ascertain how she felt. She said that she felt better. One o'clock was perfect because the weather cooled off somewhat. It was the month of June in the summertime. They decided to walk around on the first day on the park property. They saw plenty Joyney World characters walking around in the amusement park. When she saw Kinderella, Jennifer recalled her trip to Joyney World as a child. She remembered her father and mother when they were young. It was a happy part of her childhood.

15.) On Friday, they went horseback riding on a special trail. The horses are tame and know the special trail by heart. There is a guide, and the horses are trained to go slowly. The horses go slowly for safety. After the horseback ride, they rented bikes for 20 minutes on a shady path. Then, it was lunchtime. They ate outside at a quaint café under umbrellas. The outdoor café was decorated in multi-colored real flowers. The colors of the café were white and a light blue. There were fans, and the sky was a clear light blue. It was a beautiful day to enjoy the summertime weather!

16.) On Saturday, they went to the water park located inside of Joyney World. After 30 minutes of going down the waterslide, Blake and Jennifer mutually agreed to go to the lover's lagoon. The lover's lagoon is a manmade lake with waterfalls and fake mountains. It is decorated with real exotic flowers and palm trees.

17.) Sunday was just a day of rest. They went to a Catholic Church nearby to Joyney World. They were exhausted from the waterpark and swimming on Saturday. Furthermore, they both had a little bit of sunburn. They read their respective novels that they had both brought with them for an hour. Then, later they explored the three other park themes of Joyney World by walking around.

18.)   First, they went to the Epcot theme park. It contains all the futuristic rides and futuristic models. Jennifer did not want to go on any fast rides because they make her sick. To Jennifer's surprise, Blake did not like fast rides either.

19.)   Both Blake and Jennifer viewed the 'World of Tomorrow'. It had futuristic architecture. They walked through it as if it was a museum. Even after many years of having the 'World of Tomorrow', some things did become reality.

20.)   Both Blake and Jennifer toured the movie studios. There were rides that had characters from movies. It was decorated with many scenes from popular movies. There were free movies to view with two seats to view movies of one's choice.

Jennifer: Blake, could we tour the Magic Kingdom? The last time I went there was when I was a child.

Blake: Certainly, Jennifer. I would like to view the Magic Kingdom myself. It has the magic of Joyney World there. We will go there right after we are finished viewing the movie studio theme park at Joyney World.

Jennifer: Oh, how wonderful! I will try to muster up some spare energy from my adrenaline levels.

Blake: Yes, I know what you mean. It is the excitement of Joyney World! I feel it, too.

21.) Then, they walked into the Magic Kingdom after receiving stamps on their respective hands. There were cartoon characters walking around Joyney World. Jennifer spotted Kinderella and Prince Charming walking hand in hand together. Then, she saw Nicky Mouse. Donald Duck was not far behind. Kinderella's Castle was the highlight of the Magic Kingdom. Blake and Jennifer toured the inside of Kinderella's Castle. Blake and Jennifer did not like the fast rides because it makes them sick. However, they did go on the ferris wheel together. Blake put his arm around Jennifer's shoulder. They French kissed at the top on the ferris wheel. The ferris wheel stopped at the top for approximately two minutes. It was so romantic!

22.) On the way to the Luau, Blake and Jennifer saw the nighttime parade with all the lights. Kinderella was riding in her pumpkin couch. Prince Charming and other cartoon characters were in the parade, also. After an hour at the parade, Blake and Jennifer went to the Luau at night. There were hula dancers, fresh pineapples, and Hawaiian food. Blake and Jennifer just ate the fresh pineapple and salad. It was interesting to be a part of a real Luau that has all of the real customs of Hawaii.

23.) They departed Joyney World at 12 noon. Sunday afternoon. The sky was clear. It was a safe day for weather to be flying. Blake communicated with air traffic controllers to obtain a clear path to return home. Blake instructed Jennifer again to put on her seat belt. Jennifer complied with the instructions.

24.) They stopped at Tree Gardens to view the white tiger, some parrots, seals, butterflies, and other wild animals. There were some rides there, also. They stopped to have lunch there outside on a bench. They ate the most delicious dessert that was made with cheesecake, and fresh fruit. Jennifer usually does not eat cheesecake unless she makes it herself to watch out for calories and cholesterol. However, she did eat it at Tree Gardens and was extremely satisfied.

25.) Blake dropped Jennifer off at her home. He French kissed her. She responded and French kissed him back. He got her luggage off the Beep/ plane. He told her to take Monday off to catch up on her home chores and rest. "See you, Tuesday, Jennifer" said Blake.

Both Blake and Jennifer enjoyed each other's company on the vacation. The couple was respectful to each other. Respect is mandatory to have a relationship that is happy and endures. Memories were created at Joyney World, and Blake captured those memories on his high definition pocket camera.

26.) Monday morning at Cyberdroid Corporation:

Ms. Jollyday: "Someone is here to see you, sir. Her name is Dr. Charie Ballof", said Ms. Jollyday.

CEO Blake: "Show her in, Ms. Jollyday", said Blake.

Dr. Charie Ballof: Good day, sir. I am a research scientist. I have conducted many studies concerning androids. One research study firmly concluded with unbiased proof that androids are not liked in general. Many people feel that androids are taking jobs, opportunities, and other things away from human beings. Many humans feel that humans are the only ones that should have jobs, and opportunities.

CEO Blake: Thanks for your valuable time to bring this to my attention. Please email or snail mail those research conclusions to me. I would like to read them all.

Dr. Charie Ballof: Yes, sir. I shall do that the first thing tomorrow morning. I have some personal important errands to run this afternoon. Thanks, for taking the time to listen. I shall keep in touch. There is a need to have publicity directed at humans to try to create a warm and caring environment for all types of beings in this world. Everyone should have respect whether they are humans or androids.

CEO Blake: I totally agree with you, Dr. Ballof. Please contact me regularly. We will create an ad campaign to try to lessen the gap between androids and humans.

Dr. Charie Ballof: Yes, that was my sentiment exactly! I shall keep in touch. Have a wonderful day! :)

CEO Blake: Good day to you, Dr. Ballof.

27.) At Mr. Barry Benad's Private Investigator Office:

Barry Benad: (thinking…) "I do not know how I am going to bypass all the high security at Cyberdroid Corporation. Hmm… Let me see… Maybe, I have to fight fire with fire. I will utilize high tech devices to fight high tech devices. I definitely need a thermography camera that detects heat and can sense where the security guards are located in the building. Then, I need to have an entrance card, which will have to be matched to theirs. I have a wide variety of those. Maybe, not an entrance card, but a skinny mechanical saw that cuts through glass. Then, I will have to have suction cups to walk on the ceiling and bypass the motion detectors. This mission must be accomplished or I will not get paid for this job. I will search online for all new high tech devices that can help me to break in to the Cyberdroid Corporation so I can take photos of special documents concerning the creation of androids. I must not forget to wear gloves and a black outfit. The gloves are for no fingerprints. The black outfit with a mask disguises and hides me. Much planning for a successful plan is mandatory!", thought Barry.

Oh, I need an accomplice, but whom can I trust? Hmm… It would have to be some foreigner that barely speaks English. Oh, I know! That new guy at the farmer's market. What is his name? Um… His name is Racio Guerra. Yes! He will be a perfect accomplice. He needs the money.

Barry researches all new high tech devices on the Internet. Then, he goes down to the farmer's market to make a proposition to Racio. Racio accepts the offer. He is glad to be doing any kind of work.

28.) Tuesday Morning @ Cyberdroid Corporation

Blake walks to Jennifer's office inside of Cyberdroid Corporation. He knocks on the door. (Knock. Knock.)

Jennifer: Who is it?

CEO Blake: It is Blake.

Jennifer: You may enter.

CEO Blake: Did you have a good rest on Monday to recuperate from our fantastic adventures at Joyney World?

Jennifer: Yes, it was restful and peaceful. I really enjoyed our vacation to Joyney World.

CEO Blake: Wonderful! I always try to please my loved ones. There is so much more for us to do together. We will have a fun, adventurous life together.

Jennifer: Oh, Blake, you say the most romantic things to me! I adore you.

Blake: I shall call you later for lunch today.

Jennifer: I would appreciate that.

They French kiss. Then, Blake goes back to his office to do work.

29.) At Cyberdroid Corporation at night after all workers have left the building:

Barry Benad & his accomplice, Racio: They approach the Cyberdroid Corporation with the thermography camera to detect body heat of the security guards. "Racio, hand me the saw", said Barry. Barry uses the skinny mechanical saw that cuts through glass. Then, uses suction cups to walk on the ceiling and bypass the motion detectors. He enters the CEO office. He locates the hidden safe, which is behind a large oil painting in the CEO office. He attempts to open the safe. "Racio, hand me the Phillips screw driver", said Barry. "No, not the flat screw driver, you clueless idiot!", said Barry. Racio hands him the Phillips screwdriver. Within a matter of 5 minutes, the safe was opened. Barry quickly took photos of all documents concerning androids. He closed the safe, and cleaned all debris from the break. It looked as if no one broke into the safe. Barry is that good! He can do anything he sets his mind to do. He was in the Federal Bureau of Investigation (FBI). Barry and Racio walked out of the building again on the large ceiling suction cups. Then, they both ran to the get-away Kamaro small swift vehicle. "Whew! A successful break in", Barry thought. His adrenaline is overflowing. Both Barry and Racio were breathing heavily from running to the vehicle. "Ken will be so happy to see these private Cyberdroid documents about androids", Barry thought.

30.) The next morning a security guard noticed the cut glass. The police were swarming the area of the Cyberdroid Building. Blake issued a statement for the Cyberdroid workers not to come to work for 2 days. The police were dusting everything for fingerprints. They took photos of the cut glass. There were police and guard dogs everywhere inside of the Cyberdroid building.

31.) Blake went to his office with police carrying rifles in a ready position. He looked around. Everything was clean and as it should be. Nothing looked out of place. It appeared as if the safe had not been accessed. There were no valuable oil paintings missing. Everything looked perfect. Blake had nothing to report, but cut glass on the back side of the building. Blake was extremely perplexed. He thought, "What could have happened here?"

The police kept searching. They kept walking around the Cyberdroid Corporation for clues, fingerprints, broken items, stolen items, and missing items. However, there was nothing to report. The investigative team stayed there for 2 days, and still came up with nothing!

32.) After all of the investigation was over, the glass was repaired. The Cyberdroid workers returned to work. All employees were a little nervous. They were wondering if the place is safe. Some wondered if terrorists were involved. Everyone looked at his or her own personal belongings at the corporation. They had nothing to report to CEO Blake. What a baffling mystery! What could have happened? No one knows!

33.) At Cyberdroid Corporation:

Manager Adam Tyler: He knocked on the CEO Blake's office door.

CEO Blake: Come in Tyler. Blake could see him through the office door glass.

Manager Adam Tyler: Blake, all of the workers are extremely nervous. They think it has something to do with terrorism.

CEO Blake: Thanks for informing me. I will call Dr. Charie Ballof to come talk with the workers in a large group setting in the Pavilion Room. Please write an email to all office workers stating that it is mandatory to be present at this meeting. If anyone is sick, a doctor's note will be needed. Then, a private session with Dr. Ballof is needed, also.

Manager Adam Tyler: I will get right on it, sir. I will take care of everything.

CEO Blake: Thanks, Tyler. Keep up the excellent work! :)

Manager Adam Tyler departs from the CEO office.

34.) CEO Blake: (on phone) Dr. Charie Ballof, please come to Cyberdroid Corporation. I need you to speak to all of the Cyberdroid workers. Everyone is extremely nervous. They think they are not safe. They think it has something to do with terrorism.

Dr. Charie Ballof: Yes, I will come as soon as possible. Would tomorrow be fine?

CEO Blake: Yes, that is good. I appreciate your services. There will be a bonus. Would you consider being a regular member of the Cyberdroid Corporation, Dr. Ballof?

Dr. Charie Ballof: Yes, I would like that immensely.

CEO Blake: Excellent! I will have my attorney draw up a contract between us.

Dr. Charie Ballof: That is fine. However, I would like my lawyer to read the fine print before I sign a contract.

CEO Blake: That will be fine. No worries, Dr. Ballof. It will be mutually beneficial.

Dr. Charie Ballof: Ok, see you tomorrow for the speech to all workers.

CEO Blake: Thanks, Dr. Ballof, bye.

35.) The next day at the Cyberdroid Corporation:

Dr. Charie Ballof gave a speech to all of the Cyberdroid workers promptly at 10 a.m.

Dr. Charie Ballof's speech is as follows:

Good day, ladies and gentleman. It has come to my attention that all Cyberdroid office workers are extremely nervous about the broken glass. Many people think it has something to do with terrorism. I can assure you that you are in good hands here. CEO Blake has upped security to a maximum level. There are 20 more security guards stationed at entrances. New security plans are being drawn up as I speak. Every safety precaution will be taken into consideration. There is no need to worry! CEO Blake has told me that all will be fine. CEO Blake is a man of honor. I believe him, as all of you should, also. He has everyone's best interest in mind. Security is of major importance at the Cyberdroid Corporation. If anyone needs private counseling, CEO Blake is willing to pay for anyone that is in need of it. Please call me, Dr. Charie Ballof. I am in the yellow pages. Furthermore, business cards will be passed around the room. Please take one and call me at any time of the day or night. I will call you back as soon as it is possible for me to call you. I thank you for taking the time to listen. Have a wonderful, peaceful, and pleasant day!

CEO Blake: Thank you, Dr. Charie Ballof! Ladies and gentleman, please call Dr. Charie Ballof if you need counseling. I will pay for it. Everything is true that Dr. Ballof just said. We are working on security issues and plans. Please do not be alarmed. We have everything under control. Presently, we still do not know who is responsible for cutting the glass at night. However, we are trying to gather evidence. The evidence process takes time. Thanks for your time and attention. After receiving the cards, please return to your workstations to continue to execute business as usual.

CEO Blake: Thanks for your speech, Dr. Charie Ballof. You may leave now. I will call you. I must go, also. I have responsibilities in which to address.

Dr. Charie Ballof: Thanks, for calling me. I will keep in touch with you.

First, the CEO Blake and Dr. Charie Ballof exited the Pavilion room. Then, the office workers left.

36.) At Collins' Corporation:

Barry Benad: (on phone) Ken, I got some photographs of important documents that were in the safe at the Cyberdroid Corporation.

CEO Ken: I knew something was up. I saw on the news that the cops are swarming all over the Cyberdroid property like wasps.

Barry Benad: I will meet you at Chez Henri for lunch to turn the SD camera card over to you. It fits inside the laptop to view the photos.

CEO Ken: Thanks, Barry. You are terrific! There is a huge some of monetary funds that will be wired to your checking account in Switzerland. I have arranged for you to take a vacation abroad to Europe. I am sure that you need the rest and relaxation. Then, you can return when things have settled down.

Barry Benad: I will pack as soon as I get off the phone with you. Thanks, Ken. See you in one hour at Chez Henri Restaurant on Toulouse Street.

CEO Ken: Ok, bye.

The meeting at Chez Henri Restaurant only lasted ten minutes. Barry gave the SD memory card to CEO Ken. Then, he departed for the Atlanta Airport. His flight was leaving out of Atlanta in one hour. CEO Ken left as well.

37.) CEO Ken: (thinking)" I cannot wait to look at this on my laptop! I have been waiting so long for this. I am tired of being second in the information technology industry", thought Ken. Ken drove back to his office.

At Collins' Corporation:

CEO Ken put the SD memory card into his laptop as soon as possible. There were schematic drawings, diagrams, and formulas. None of it appeared to make any sense. It was written in some sort of code for security purposes. "Damn it! It appears Blake outsmarted me again!", exclaimed Ken. Ken hit his desk with his fist. "If it is the last thing that I ever do, I will succeed over Blake Fencington and the Cyberdroid Corporation!", said Ken.

CEO Ken kept his word to private investigator Barry. One million dollars was wired to Barry's personal checking account in Switzerland. Ken thought, "Barry, did his best, and retrieved what I asked. Maybe, one of the engineers here at Collins' Corporation can decipher this. It is worth a try", thought Ken.

38.) CEO Ken: John, please come to my office now. I need to speak with you. It is important.

John: Yes, Ken. I am coming now.

CEO Ken: Can you make anything out of this?

John: No, it is some sort of code. I never saw it before. It must be a personal code because I know of many codes and this does not look like any of them that I have ever seen.

CEO Ken: Drats! What rotten luck! Nothing, absolutely nothing. I have a formulas and secrets, and yet I cannot decipher the code. I wonder if anyone else here knows anything.

John: Well, Ben might know something, but that is it. All the others are not as knowledgeable.

39.) CEO Ken: (on phone) Ben, please come to my office at once.

Ben: No, Ken, I am sorry. I do not recognize that kind of code.

CEO Ken: Oh, man! Now, what will I do? Thanks, gentleman for your efforts. I appreciate it. I need to be alone to think now.

John and Ben left the CEO office and went back to their office stations.

40.) At Cyberdroid Corporation:

CEO Blake: Jennifer, please come into my office.

Jennifer: What is on your mind, Blake?

CEO Blake: It has been a few months since the break in. Things have settled down. I really need a vacation! How about us going to vacation at a log cabin in Northern Georgia. It is beautiful this time of year in the fall with the leaves turning red, orange, and yellow.

Jennifer: Yes, I love the fall, also. I will go home to pack when I get off of work.

CEO Blake: Does tomorrow at 10 a.m. sound fine to leave for our vacation?

Jennifer: Yes, do not worry. I will be ready.

CEO Blake: Excellent. You have a lovely day, Jennifer. See you tomorrow.

Jennifer walked back to her office station at Cyberdroid Corporation.

41.) CEO Blake: (on office intercom) Oh, Ms. Jollyday I just received an email from former President Strong-tree. He heard of the break-in on the national and world news on channel seven with reporter Benson. He would like to come give a speech to the office workers. He said that he knows that they must be upset especially since a few of the Cyberdroid employees told the news that they thought that terrorists were responsible for the break-in.

Ms. Jollyday: (on speakerphone) Oh, how lovely a visit from former President Strong-tree. I always admired that President of the United States of America. I always thought that he was so brave and dedicated to Mr. Ban Leaders' capture. I felt relieved after hearing President Strong-tree's speeches on television. He was one of the best Presidents of the United States of America! I will make the necessary arrangements by writing an office memorandum email to all Cyberdroid workers to be present at President Strong-tree's speech here at the Cyberdroid Corporation in the Pavilion Room.

CEO Blake: Thanks, Ms. Jollyday. Please order some catering, refreshments, and soft drinks for the event. You are an efficient worker. I will call President Strong-tree to tell him to come to the Cyberdroid Corporation at his earliest convenience.

Ms. Jollyday: Oh, that is fantastic!

The next day, CEO Blake walks into the front office.

42.) CEO Blake: Ms. Jollyday, I am leaving for Beijing, China in 20 minutes on Cyberdroid's Supersonic Company Jet. I must go view the stronger security measures that were implemented at the android manufacturing plant in China. I will be back in 4 hours.

43.) CEO Blake departs on the Supersonic Company Jet. He arrives in Beijing, China in one hour. Blake is greeted by Mr. Glenn Ganuchi. Mr. Ganuchi is lead scientist at the android plant in Beijing, China.

CEO Blake: Oh, Mr. Ganuchi, it is so good to see you! You are doing a wonderful job here at Cyberdroid's android manufacturing facility.

Mr. Ganuchi: (He has a Chinese accent.) Thank you, Mr. Fencington. I take my job extremely seriously!

CEO Blake and Mr. Ganuchi are walking down the long corridor. They enter the room containing the androids. The androids are located in long individual clean glass cylinders. They are standing up in the glass cylinders completely nude. They look like full-grown people. There are all kinds of stainless steel machines all around the cylinders for life support systems. The androids are in the last phase of creation. They can actually breathe on their own, and have all regular bodily functions like human beings.

44.) CEO Blake went up close to one of the cylindrical glass containers containing a blond female android. She could actually open and close her eyes. She heard them talking and opened her eyes to look at them.

CEO Blake: She is beautiful!

Mr. Ganuchi: She is intelligent, also. I am now transferring data into her mechanical/human brain. As you already know, androids have part human and part mechanical brains.

CEO Blake: Yes, it is just an amazing process! I am so glad that I received divine communication from the Creator of the Universe.

Mr. Ganuchi: I agree. Androids have much good to contribute to the world.

CEO Blake: I second that.

Mr. Ganuchi: They have contributed good and beneficial things already so far. I keep up with the news. At least, the ones who want to reveal their identities as androids. Other than that, no one could ever really know that they are androids.

CEO Blake: Mr. Ganuchi, you have done an excellent job implementing the new and innovative security procedures and devices. I can always count on you! You have a large bonus coming to you at Christmas time.

Mr. Ganuchi: Thanks, Mr. Fencington. It is a pleasure working for you.

CEO Blake: I must get back to Atlanta, Georgia, United States of America now. I have a meeting with a client at 4 p.m. Eastern Time Zone.

45.) CEO Blake made it back to Atlanta, Georgia at 3:50 p.m. President Strong-tree arrived at 4 p.m.

CEO Blake: Oh, President Strong-tree, it is such an honor to see you again.

President Strong-tree: Likewise. I always enjoy our visits.

CEO Blake and President Strong Tree shake hands and smile warmly at each other.

CEO Blake: How have you been?

President Strong-tree: Well, I keep busy with the grandchildren from both of my daughters.

CEO Blake: Yes, it is a joy to have grandchildren. Children bring so much joy and happiness to the world's population.

President Strong-tree: Yes, I agree with you, Blake.

CEO Blake: You can freshen up in my personal bathroom before leaving. Then, I will escort you to the Pavilion Room. Would you like some spring water?

President Strong-tree: No, I am fine. Barelle and I just finished eating some ice cream before we came here.

CEO Blake: Are you ready to give your speech to the Cyberdroid workers, President Strong-tree?

President Strong-tree: Yes, I am ready. I truly want to make the Cyberdroid workers feel at ease since the break-in.

CEO Blake: Fabulous! I will let the workers know by intercom to go to the Pavilion Room now. You can give your speech in approximately 20 minutes.

46.) After approximately 20 minutes, President Strong-tree and CEO Blake walked down the corridor to the Pavilion Room. As usual, President Strong-tree was surrounded by six security guards. President Strong-tree, Mrs. Barelle, CEO Blake, and the six security guards entered the Pavilion Room at the Cyberdroid Corporation.

CEO Blake: Ladies and Gentlemen, may I have your complete attention, please. I have the proud pleasure of introducing President Strong-tree. President Strong-tree came here to provide us with some words of wisdom.

President Strong-tree: Thanks, Blake.

47.) President Strong-tree's Speech:

Ladies and gentlemen, I heard on the national and world news about the break-in. I wanted to come talk with all of you. Please do not be afraid. We, as Americans, should not live in fear. I promised you that I would capture Mr. Ban Leaders and I did. I set up the arrangements with military leaders on the plan to capture Mr. Ban Leaders. It took a few years. However, justice was served. You as office workers need to be strong and fearless. Do not let terrorists stop you from your work and hobbies you love to do. Live life to the fullest! You are in good hands with Blake. I trust him to execute a terrific and safe security plan. You must do the same. Be strong, proud, and brave Americans! Thank you for your time and attention.

A standing ovation occurred from all of the Cyberdroid office personnel, CEO Blake, Barelle, and the six security guards.

CEO Blake: Thank you, President Strong-tree. I have enjoyed your speech.

President Strong-tree: Blake, I have to go now. I have an appointment early in the morning tomorrow. I will call you to keep in touch and see how everything is going. Maybe, you can take a lady friend to one of my summertime family barbecues.

CEO Blake: Thank you for coming, President Strong-tree. I certainly will take you up on your summertime family barbecue. I will call you to keep in touch.

President Strongtree, Mrs. Barelle, and the six security men left the building.

48.) CEO Blake: Ladies and Gentlemen, you are free to leave early today. Go home to relax and spend some quality time with your respective families and children. See you one week from today. I am giving everyone an extra paid week vacation to recuperate and have some free time.

There was a standing ovation again, but this time it was for CEO Blake. Blake left the Pavilion Room before the Cyberdroid workers left.

49.)   The next day, CEO Blake receives a judgment which was hand delivered by a policeman. The policeman brought it directly to Blake at his CEO office. Ms. Jollyday was shocked. She sensed something was wrong, but she did not know what.

Ms. Jollyday: What is it all about, Blake? Is the Cyberdroid Corporation being sued?

CEO Blake: Yes, Ms. Jollyday, I am afraid so. It appears we are being sued.

Ms. Jollyday: By whom? Who would do such a thing? The Cyberdroid Corporation pays taxes and abides by the laws.

CEO Blake: It is from the top competitor, the Collins Corporation. Ken Collins is behind all of this.

Ms. Jollyday: But, why? What did we ever do to him?

CEO Blake: Nothing, Ms. Jollyday. That is just it, nothing.

Ms. Jollyday: Well, what grounds does he have to sue?

CEO Blake: Give me a few moments, Ms. Jollyday, and I will tell you.

Ms. Jollyday: Ok, sorry.

CEO Blake: After reading through the first few pages, CEO Blake has a perplexed look on his face.

Ms. Jollyday: I never would have thought in my wildest dreams that this would occur. It appears that Ken Collins is suing the Cyberdroid Corporation for having a monopoly on the manufacturing of androids.

Ms. Jollyday: Oh, that is absurd! How could he do that? You applied for patents, copyrights, and trademarks for anything dealing with androids or the Cyberdroid Corporation.

CEO Blake: Well, he will have his day in court.

Ms. Jollyday: Do not worry, Blake. Everything will be fine. The judge knows you are a fine citizen and give back to society. You abide by all the rules and regulations concerning corporations. You pay your taxes on time. Do not worry, Blake.

CEO Blake: Well, I know Ms. Jollyday. However, it is still disturbing to me. I have never been to court in all of my life.

Ms. Jollyday: Yes, it is disturbing, but I am looking on the bright side of things. Have faith. I will pray for you and the Cyberdroid Corporation.

CEO Blake: Thanks, Ms. Jollyday. You always make me feel better.

Ms. Jollyday: You are welcome.

CEO Blake: Ms. Jollyday, you have a good evening. I am leaving now. I need to relax, and rest my nerves.

Ms. Jollyday: Yes, I understand. I will lock up. There are only fifteen more minutes to closing time.

50.) The next day, CEO Blake, visits Jennifer in her office at the Cyberdroid Corporation.

CEO Blake: Jennifer, I need to take a short weekend trip. We can go to the north Georgia mountains. Then, we can go white water rafting on the Nantahala River. It is only a class 4 classification for danger.

Jennifer: Ok, it does not take me long to pack. In fact, I keep two suitcases packed. One has fall clothes in it with accessories. The other suitcase has spring clothes in it. Therefore, I am ready in a minute's notice.

CEO Blake: We will drive there on the scenic route. Is that fine with you, Jennifer?

Jennifer: Yes, it is fine with me.

CEO Blake: I love to see the different colors of the changing leaves. The many shades of yellow, orange, green, tan, and brown are lovely at this time of year in the fall.

Jennifer: Yes, my two favorite times of the year are spring and fall. The weather is not too hot, nor too cold.

CEO Blake: Yes, I agree with you, Jennifer.

51.) CEO Blake and Jennifer departed from her house. They rode along I – 75 from Atlanta, Georgia to Chattanooga, Tennessee. Then, Blake drove up to Cleveland, Tennessee.

They had planned to go white water rafting on the Nantahala River, which is a class 2 and is not dangerous. The trip would be in the log cabin in the mountains of Gatlinburg, Tennessee. There was a log cabin there that Blake owned, which had all the amenities of being at home. It even had a heated swimming pool, sauna, and hot tub. It was surrounded by plenty of vacant land that had trees on it. Blake had purchased a large plat of land surrounding the log cabin for privacy and peace. The log cabin was designed by Blake himself. He had been the contractor for his log cabin.

The log cabin was beautifully decorated in a shade of medium blue. The sofa and chairs were medium blue. The china was white with medium blue on the sides of the plates. There was a second floor of bedrooms that overlooked the living room. There were plenty large windows all around the log house to see the maintained gardens with different color rose bushes. The log cabin had a fireplace with real logs burning in it. The log cabin was so large that it had a retired elderly couple that lived in one section of it to maintain the land and to keep the interior clean. However, at the time of Blake and Jennifer's vacation, the elderly couple had taken their vacation to Paris, France to visit some family members. Therefore, Blake and Jennifer were totally alone and felt at ease being themselves.

52.) Blake: Jennifer, we are finally here. Isn't it beautiful?

Jennifer: Yes, it is beautiful.

Blake: I designed it totally myself, and was the contractor when it was built.

Jennifer: Yes, I like the architectural design.

Blake: No one is here. The elderly couple that maintains the log cabin and surrounding grounds are on a vacation to Paris, France to visit some family members.

Jennifer: Oh, that is nice.

Blake: That is wonderful! We can enjoy the time alone as a couple.

Blake pulled up to the front of the log cabin and got the suitcases out of the back of the Beep Super Utility Vehicle. Then, he went around to the other side to open the door to help Jennifer out of the vehicle. They walked to the front of the log cabin. He unlocked the door and turned the burglar alarm off. Then, he picked up Jennifer to carry her over the threshold of the doorway.

Jennifer: (laughing) Oh, my goodness! This is so romantic!

Blake: Yes, I want it to be romantic. You are so special to me! We have been dating a couple of years now. I feel it is time now.

Jennifer: Time for what?

Blake got down on his knee. He held open one of the most beautiful rings in the entire world. It was a sapphire medium blue shiny stone to match Jennifer's eyes. The sapphire stone was in the shape of a heart. The heart stone was two carets large with some small shiny clear diamonds surrounding it. The band was a pure platinum metal. It glistened in the sunlight.

Blake: Jennifer, will you marry me?

Jennifer: I would be honored to be Mrs. Blake Fencington. You have always treated me with respect and love. You are a kind, gentle, and soft-spoken, respectful man. I love and adore you!

Blake: Thanks, Jennifer. You will not regret it. I will try to make every day perfect for you.

Jennifer: I will try to do the same. All I ask is that we keep the lines of communication open for anything that is bothering us. Communication is the key to any good relationship. However, topics must be approached in a respectful manner to the other person. There should be no screaming, or yelling of any kind. Furthermore, we must always try to meet each other half way on topics in which we do not agree.

Blake: That is a deal. I will try my best to make you happy, Jennifer.

Jennifer: Thanks, that is all I ask. Trying is half the battle towards a successful relationship.

Blake: Jennifer, I saw you yawn a few times. Let me show you to your room so that you can rest. We can spend quality time together tomorrow. Have a good night's sleep. Goodnight, Jennifer.

Jennifer: Goodnight, Blake.

53.)   Blake French kissed Jennifer on the lips while hugging her. Then, he left the bedroom, and went into the bedroom adjacent to Jennifer's bedroom. Both were tired from viewing the scenic route with the beautiful fall foliage of yellow and orange leaves. It was about 8 p.m. at night. They both took a shower simultaneously, but in separate bathrooms located in each respective bedroom. Then, they both went to sleep early at approximately 8 p.m. at night.

54.)   The next morning Jennifer heard some noise in the kitchen. It sounded like a coffee grinder. Then, there was another noise of coffee percolating. Jennifer wanted to look beautiful for Blake. She flossed and brushed her teeth. Jennifer likes to use the new fluoride rinse that refreshes breath, also. She shaved her under arms, and legs. Then, she urinated in the toilet. Jennifer soaped up with generic Love Bird Liquid Soap. She rinsed the soap off and got out of the shower. Jennifer then lightly creamed her whole body with cucumber/melon cream. She dressed in a dark blue dress that fitted tightly around her breasts and waist. The bottom of the dress was ruffled to knee length. Her hair was blown dry with a hand dryer on a warm setting. Jennifer pulled one side of it up. Then, she secured it with an ornamental heart shiny barrette. The other side was flipped back. Jennifer always has a side part in her lovely strawberry blond, silky long hair. Her shoes were light tan sandals that buckled at the ankles. She looked so beautiful!

55.) Blake: Oh, my, Jennifer, you look so beautiful! I love that dark blue dress on you with your golden hair flowing softly down.

Jennifer: Thanks, Blake. (She kissed him on the cheek, and put her arm around his shoulder.)

Blake: Sit right here my princess. I have cooked a delicious and nutritious breakfast for us.

Jennifer: (She sat down in the designated chair.) Oh, it all smells so wonderful! What did you cook?

Blake: Well, I made a breakfast sandwich for each of us. It is egg whites only on whole wheat bread with a slice of 98% lean turkey, and a slice of 2% milk Colby-Jack cheese melted in the microwave. I brewed some freshly ground coffee for us, also.

Jennifer: I did not know that you knew how to cook a delicious and nutritious breakfast for us.

Blake: I am a man of many talents; as you will learn over the course of time.

Jennifer: Yes, I can see that.

They ate their respective healthy low fat, whole-wheat sandwiches. Then, Blake put the dirty dishes in the dishwasher. After they ate, Blake asked Jennifer to take a walk on the grounds surrounding the log cabin. The gardens surrounding the log cabin were beautiful. It was still warm enough to have flowers. The rose blooms were in full bloom and plentiful. It smelled like perfume when they passed the clusters of rose bushes that were trimmed neatly.

Blake: Let us brush our teeth again swiftly in our respective bathrooms so we can hurry. Then, we will take a walk.

Jennifer: Ok, I always like to take care of my teeth.

Blake: Are you finished, Jennifer?

Jennifer: Yes, I am just applying some sunscreen, and lip balm with SPF 15. I will be out in a minute.

Blake: Ok, Jennifer.

56.)   Blake and Jennifer held hands while walking on the grounds around the log cabin. It was about 10 o' clock in the morning. The wind was blowing softly through Jennifer's long, silky golden hair as they were walking outside under the intermittent shady trees and rays of golden sunlight. The wind then made a 'swoosh' sound and rustling noise in the fallen multi-colored yellow, orange, red, and tan leaves. There were pine trees, red maple trees, silver maple trees, peach trees, cherry trees, plum trees, and a red seedless grape vine. It was an ideal place to chat about different topics that come to mind along the peaceful and serene walk. Sometimes, it was just fine being quiet to enjoy the peaceful, serene aspects of nature at its best.

57.)   Blake and Jennifer finally finished touring the land around the log cabin. They decided to sit in a gazebo. It was located about 10 feet away from a 2-foot wide water stream that came directly from the higher part of the Smokey Mountains. The 2-foot wide water stream ran down the side of Blake's land almost on the boundary line of the next plat of land on the far end of the log cabin. That was perfect because Blake did not want any flooding on his land near the log cabin, especially in the springtime when the snow and ice melts.

Blake stroked Jennifer's hair out of her eyes.

Blake: Jennifer, I truly adore you. (Blake gazed into Jennifer's eyes as he said that.)

Jennifer: Blake, I feel the same about you!

Then, Blake French kissed Jennifer tenderly on the lips and hugged her. They sat close to each other for an hour on a bench swing in the white gazebo that had many different kinds of flowers planted around it. There were yellow daisies, red roses, purple hydrangeas, white daisies, white roses, lavender roses, and blue forget-me-nots. It smelled aromatic like perfume. However, the smell was not too strong. It was just right. It was so relaxing in the white gazebo. They almost fell asleep because the temperature was perfect. It was not too hot and not too cold in the fall. Blake and Jennifer loved the fall and spring the best because of the mild temperatures, fall foliage, and the last of the summer flowers. However, roses are wonderful because they are perennials that last all year long with budding roses even in the wintertime.

58.) Blake: Jennifer, let's go back to the log cabin because it is lunchtime. I make the best 98% lean turkey breast sandwich on whole-wheat toasted bread. I, also, make delicious freshly brewed green tea. Green tea has plenty anti-oxidants for health benefits.

Jennifer: Ok, I try to eat right, and exercise regularly.

Blake: I try to eat healthy and exercise regularly, also.

Blake made the lean turkey sandwiches on toasted whole wheat bread. Then, he brewed some fresh green tea. He added ice cubes, some aspartame sweetener, and lemon slices. It was a delicious and nutritious lunch. For dessert, they ate 4 ounces of low-fat chocolate ice cream with marshmallow cream and pecans.

59.) Blake: Jennifer, I have this DVD movie of Gim Karey, I just love Gim Karey, don't you?

He is so hilarious sometimes. He is the best comedian! It is called, 'No, man'.

Jennifer: Yes, I heard about it. It is supposed to be hilarious! I love Gim Karey, also!

Blake: Ok, I will get it from my collection. Just a moment, please.

Jennifer: Ok, take your time. I will make us some fresh air-popped popcorn. Then, I will spray it lightly with spring water so the salt will hold onto the popcorn. It is a fat free snack.

Blake: Ok, sounds terrific!

They watched the DVD movie of Gim Karey. Both Blake and Jennifer enjoyed it immensely. They both laughed excessively. It is a good emotional relief from stress.

Blake: I am kind of tired. I think we should take a short one-hour nap.

Jennifer: Ok, I need the extra rest myself.

59.)   They fell asleep on the medium blue carpet in front of the high definition television, which was turned off after they viewed the movie. Blake set one of his many alarm clocks that are located in different areas of the log cabin. Blake and Jennifer used the sofa pillows and a medium blue blanket on the sofa. They took a nap in each other's arms. The log cabin has a central air and heat, which stays at 79 degrees. It is not too hot, nor too cold. That makes an ideal setting to fall asleep for a nap.

60.)   When the alarm went off, it was 4 o'clock. Both awakened in each other's arms.

Jennifer: I will go brush my teeth to be fresh for you.

Blake: Excellent idea! I will do the same.

They both returned to the living room in the log cabin.

Blake: What shall we do now? How about a dip in the hot tub? I give wonderful back massages.

Jennifer: Ok, good thing I have my bikini packed.

Blake: Let's use the bathroom, and then put on our respective bathing suits.

Jennifer: Ok, that sounds fine.

They changed into the bathing suits, and met in the living room. Then, Blake escorted Jennifer to the hot tub. They both got into the hot tub simultaneously. The warm, bubbling water felt so good and exhilarating! Blake and Jennifer chatted and took turns massaging each other's backs. They stayed in the hot tub for 1 hour until it was time for dinner. They both got out and wrapped the large thick dark green clean towels that Blake got before entering the hot tub. They both showered and changed for dinner.

Blake: Jennifer, be sure to wear formal attire. I like to dress for dinner.

Jennifer: Ok, Blake. I will put on my best outfit to try to look beautiful for you.

Blake: You are always beautiful, Jennifer. (Then, Blake hugged and kissed Jennifer on the cheek.)

They went to their respective bedrooms with adjoining bathrooms to shower and get ready for dinner together.

61.) Blake was waiting for Jennifer in the living room of the log cabin. He was dressed in tan dress pants, and a tan and gold print polo shirt. Jennifer emerged from her room an hour later. She had brushed her teeth, shaved her arms and legs; washed and conditioned her hair; and blow- dried her hair. Then she lightly applied sunscreen, blush, lip-gloss, and light blue/ grey eye shadow. Jennifer dressed in a dark green and white print sweater that fit tightly around her breasts, arms, and waist. It was fall. At night, it was starting to get a little brisk and chilly outside. The temperatures dropped down to the sixties that night. Her pants were a dark green corduroy material with side pockets and elastic around the waist. Jennifer styled her hair down and combed to the side. It looked lovely with her heart-shaped barrettes.

62.) Blake had dinner catered from one of a chain of the Chez Henri French Restaurants. The dinner was freshly grilled salmon, sautéed thin green beans, and fresh circular-cut potatoes. The doorbell rang. The delivery driver was prompt and courteous. He delivered dinner at 5:50 p.m. Blake took the two dinners; paid the delivery driver, and even gave him a 15% gratuity tip for the food. The food smelled wonderful, and made Blake want to eat it as soon as Jennifer was ready to eat.

Jennifer came to dinner at 6 p.m. She looked beautiful as usual. She sat down at her place setting. Blake had set the table elaborately with a white lace tablecloth and medium solid blue napkins. He even knew that the knife was on the left, the spoon in the middle, and then fork on the right. Jennifer was impressed.

Blake: I hope it is to your satisfaction. I had the dinners catered from Chez Henri, the famous French restaurant chain.

Jennifer: Yes, it smells delicious!

Blake: (Blake made the sign of the cross.) "In the name of the father, the son, and the holy spirit."

Blake and Jennifer: "Bless us oh, Lord, in these thy gifts, which we are about to receive from thy bounty through Christ our Lord, Amen."

They both made the sign of the cross. Then, they both ate their food.

Blake: Jennifer, I admire your beauty. No matter how many times I see you, you still look like the most beautiful lady in the world to me.

Jennifer: Thanks, Blake. I appreciate the fact that you appreciate me! I admire your handsomeness.

Blake: Thanks, Jennifer. (He leaned over to kiss Jennifer on the cheek.)

Dinner went well. Both Blake and Jennifer love dinners catered from the famous French Restaurant chain called Chez Henri. After dinner, they sat on the sofa together to watch the World evening news with Ryan Jenson.

Jennifer: I love to watch, Ryan Jennson. He is so informative, and delivers the news as truthfully as possible.

Blake: Yes, world news has to be reliable.

Jennifer: Oh, that is terrific! No rain for the remainder of our vacation here in Gatlinburg, Tennessee.

Blake: Yes, that is wonderful news!

Blake and Jennifer were both tired and went to sleep early that evening.

63.)   The next morning, Blake and Jennifer, had freshly brewed coffee, low-fat cake, and their multi-vitamins for breakfast. After breakfast, both Blake and Jennifer exercised in the exercise room, which had many different kinds of exercise equipment. Blake and Jennifer rode indoor aerobics bikes side-by-side simultaneously for 20 minutes each. Jennifer likes to rest for at least 10 minutes before going to lift arm and leg weights. Blake was flexible and did what Jennifer did. They both rested for 10 minutes before lifting weights. Blake lifted 100 repetitions of arm weights and 100 leg weights. Jennifer just lifts 50 arm weights and 50 leg weights.

64.)   At 10 o' clock in the morning, Blake and Jennifer went to their respective bathrooms to shower and get dressed for the day ahead. They were both sweaty from exercising even though the log cabin is fully air-conditioned. They both washed their hair. Jennifer has the added task of shaving her legs.

Jennifer wanted to look gorgeous for Blake. She took her time to make certain that she could look as beautiful as possible. She used the bathroom first. Then, Jennifer shaved her legs. She always washes her hair before her body so the creme rinse can condition the hair by the time she is finished bathing. Jennifer loves the Australian shampoo and creme rinse. It works well and smells like candy.

Jennifer dressed in a medium blue dress with a long sleeve tight fitting sweater underneath. It was tight fitting at the breast, waist, and arms. The bottom of the dress reached the knee and looked puffy. Her hair was freshly washed, and flowing softly down around her face. She looked like a beautiful doll in her dress.

65.) Blake: (He was waiting for her in the living room of the log cabin as usual. This time he was reading a science-fiction novel while he was waiting.) Oh, Jennifer, you look so beautiful!

Jennifer: Thanks, Blake. I could hear you tell me that for the rest of my life.

Blake: I want to tell you that for the rest of our lives together as a couple.

Jennifer: Oh, Blake, you are so romantic!

Blake: Would you like to go walking down Main Street in Gatlinburg, Tennessee? The shops are so colorfully decorated like a quaint little German town.

Jennifer: Yes, that is not far. It is just at the foot of this mountain.

Blake: Ok, let us tour Main Street.

Jennifer: That is fine with me.

Blake: I will pull the Beep SUV around to the front of the house. You can wait on the porch.

Jennifer: Ok.

66.) Blake turned off all of the lights, and left on the kitchen light. He, also, left a few night-lights on around the house. It would be dark when they returned. Then, he turned on the burglar alarm. He pulled the Beep SUV vehicle around the front of the log cabin. Jennifer got into the vehicle on the passenger side. Blake drove down the steep side of the mountain to the German-looking shops below.

67.) Blake parked at the base of the mountain. His log cabin is located right above Main Street in Gatlinburg, Tennessee. He did not want to get towed away so he parked in a parking garage nearby. They walked down Main Street together. Blake had his arm around Jennifer's shoulders for a short while. Then, they strolled down Main Street holding hands while walking together. They went into numerous shops all along Main Street. They just browsed at the tourist merchandise for sale. The shops were the following: candy shops, craft shops, clothing shops, and homemade Indian Jewelry shops. They passed some restaurants along the way. At 4 p.m., they both agreed to eat at the Trout Charlie Lee. The specialty there was fresh trout. The trout swim up streams all along the mountain to spawn their fish eggs. The trout is so fresh and delicious! Both Blake and Jennifer enjoyed their grilled trout dinners with baked potato, and salad on the side. After dinner, they walked back to the parking garage to retrieve the Beep SUV. Blake paid the parking fee, which was $ 5.00. Then, Blake drove them back to his log cabin, which was located right above Main Street in the mountains of Gatlinburg, Tennessee.

68.) Blake and Jennifer watched a DVD movie. It was called, 'Mr. Bopper's Penguins'. It was about how the main character's father sends penguins to his son's house. It was an extremely informative movie about penguins. The movie was enjoyable to watch, also.

Blake: I just love Gim Karey. He is so hilarious!

Jennifer: Yes, he is one of my favorite actors, also.

69.) Both Blake and Jennifer watched the DVD movie. It was about 6 o'clock when the movie was over. Blake and Jennifer started to kiss and hug passionately for about 30 minutes. Then, they both decided to give each other invigorating back massages. At 7 p.m., both Blake and Jennifer were extremely tired. They went to their respective bedrooms to sleep. Their romantic fall vacation was over. The following day, they had to return back to Atlanta, Georgia. Monday would be spent at their respective houses. Then, Tuesday would be the day to return back to work.

70.) After returning from the short vacation with Jennifer, CEO Blake had to go to court. He left in his Beep SUV/plane. His lawyer, Chris Murphy, met him at the courthouse. Both left simultaneously.

CEO Blake and his lawyer Christopher Murphy arrive at the Courthouse. They walk through the parking lot building together. Then, they go in through the glass doors. There were security guards on each side of the metal-detector-walk-through stations. They walked down some corridors to a courtroom. CEO Blake and attorney Chris took a seat in the front row on the left side of the courtroom, which is designated as the defendant side. Everyone in the courtroom was chatting softly. The courtroom filled up fast because an announcement was made by the Collins Corporation. Reporters were there interviewing people in the hall. No video cameras, or cameras of any kind are allowed in the courtroom. The bailiff, Joe Jackson, stood in front of the courtroom.

Bailiff Joe Jackson: Order in the court. The honorable Judge Jack Murphy is presiding. Everyone please rise.

Everyone became quiet. It was so quiet that it was like a library. Everyone rose in respect of Judge Jack Murphy.

Bailiff Joe: You may have a seat now.

The judge walked over to his chair, which was elevated about 3 ft. from the floor. It was located in the center front of the courtroom. The courtroom was decorated elaborately with wood molding in an oak light tan wood color.

Judge Jack Murphy: Everyone should keep quiet while court is in session. Whispering is allowed if it is truly needed for court proceedings. Discussions between client and lawyer should be conducted in the small discussion cubicles located on each side of the corridor. Cell phones must be turned off at all times. Cell phones can be put on silent vibration to receive emergency calls. However, calls must be taken in the hall away from doors in a low voice. There must be absolutely no type of video cameras or cameras of any kind. Do not force me to have to reprimand you. I will take the cell phone away until the end of the daily court session. Let us proceed with the matter at hand.

71.) Cyberdroid's lawyer, Mr. Chris Landry: Your honor, we have been served with the court papers that state that the Cyberdroid Corporation has a monopoly on android creation and manufacturing. It is totally ridiculous and I say this because everyone in the United States of America has rights and freedom. One of those freedoms is to apply and obtain copyrights, patents, and trademarks. My client, Dr. Blake Fencington, did apply for all mandatory copyrights, patents, and trademarks. He did everything according to the law. Dr. Blake Fencington pays his taxes and is a philanthropist. He donates brand new desktop computers to needy schools, colleges, universities, and libraries. Dr. Blake Fencington is an honorable member of society. He should have the respect, rights, and freedoms that each and every citizen of the United States of American has. Furthermore, Dr. Blake Fencington solely created everything he ever produced, manufactured, or sold. Of course, there are helpers who are scientists who work for Dr. Blake Fencington. However, all work was conducted for the Cyberdroid Corporation. Everyone signed and agreed on contracts. All was done according to the laws of this state of Georgia in the United States of America. I implore you, Judge Murphy, to take all of this into consideration. Please give Dr. Blake Fencington the respect that he deserves as a commendable citizen of society. (Attorney Chris Landry stepped down from the podium which was located in the center of the courtroom.)

72.)   An attorney from the plaintiff side walked before Judge Jack Murphy to the center of the courtroom. The plaintiff's attorney, Bill Walker, approached the Judge.

Collins Corporation Lawyer, Bill Walker: Your honor, my client, Ken Collins is Chief Executive Officer of Collins Corporation. Collins Corporation manufactures desktop computers and laptops. He feels that there is a monopoly on android creation at the Cyberdroid Corporation. If the blueprints and instructions were made public, then anyone could create androids. Therefore, the prices would fall. It is for this reason that I implore you to make the secrets on android creation available to everyone in society. Thanks, for taking the time to listen.

Judge Jack Murphy: There will be a one-hour recess. Everyone can take a break now. I will think about what was said from both sides of this case. Everyone should return to this courtroom one hour from now.

The judge and bailiff left the courtroom. Everyone started chatting. Some people left the courtroom. A few people stayed to chat softly.

73.) After one hour, everyone returned to the courtroom.

Bailiff Joe: Everyone please be quiet. Please stand for the honorable Judge Jack Murphy. Court is now in session. Please turn off your cell phones, or I will confiscate them until the end of trial.

Judge Jack Murphy: I have listened to both sides of this case. I have reached a verdict. Android creation and manufacturing is an entirely new field. It is for this reason that there are no preceding laws to use for this case. However, I have to base my decision on the 'Bill of Rights of the United States Constitution of America. It is for this reason that I have to agree with the defendant side, the Cyberdroid Corporation. Everything that attorney Chris Murphy said is true. Dr. Blake Fencington applied for copyrights, patents, and trademarks. He is a respected citizen of society. He must be given the same rights and freedoms that all other citizens of the United States of America are given. Therefore, I rule in favor of the Cyberdroid Corporation and Dr. Blake Fencington. Dr. Blake Fencington does not have to reveal his secrets on android creation and manufacturing to the general public.

74.) Collins Corporation Bill Walker: I object! We will appeal!

Bailiff Joe: Order in the court! Everyone please be seated and remain quiet! All people in court were chatting now.

75.) Judge Jack Murphy: He left the courtroom through the side door.

76.) Bailiff Joe: Court has ended. Please exit the courtroom in an orderly fashion now. Anyone that gives me trouble will be arrested. Please leave now.

   Bailiff Joe got on his cell phone. He called the police.

Bailiff Joe: I need some police here at the Atlanta, Georgia Courthouse immediately.

   Everyone left the courtroom slowly still chatting about the verdict.

77.) At Collins Corporation:

CEO Ken: (thinking about the court proceedings…Ken slammed his fist down on the desk.) "Dammit, it could have worked! I was so close, and yet so far…", said CEO Ken. Then, Ken rubbed his hand that hit his desk.

CEO Ken: (on cell phone) Chrissie, please come into my office.

Chrissie: Yes, Ken. I am coming.

   Chrissie walked down the hall to Ken's office. She knocked on Ken's office door.

CEO Ken: Chrissie, I cannot believe the Judge. He made up his mind quickly. I even said that it was a monopoly. Yet, it did not work.

Chrissie: Oh, I am sorry that it did not work out for you, Ken. Just look on the bright side. He could have made everyone go to the court for 2 weeks, and still had the same decision. This way no one lost as much time. Time is valuable. Time is money in business.

CEO Ken: Yes, I suppose so. Still, I have a bad feeling.

Chrissie: Yes, I know what you mean.

CEO Ken: Thanks, Chrissie, I appreciate your opinion. Would you like to go to lunch with me at 12 noon tomorrow?

Chrissie: Yes, thank you.

CEO Ken: Ok, meet me here at my office tomorrow for 11:30 a.m.

Chrissie: That will be fine.

78.) Chrissie left CEO Ken's office.

CEO Ken: Barry, how have you been?

Former FBI Agent Barry Benad: Fine, and you, Ken?

CEO Ken: I have had the worst, rotten luck, Barry!

Barry Benad: What happened, Ken?

CEO Ken: Well, you read the newspaper and watch the news. I am sure you must have seen something about it.

Barry Benad: Yes, Ken. I did see it on the World News Tonight with Ryan Jennson. You are referring to the Judge's decision on your case, aren't you?

CEO Ken: Yes, Barry, that rotten luck.

Barry Benad: Do not worry about it. Your time will come.

CEO Ken: Yes, but when? I have been second in the information technology industry for quite a long time now.

Barry Benad: Is there something I can do for you, Ken?

CEO Ken: Yes, now that you mention it. There is something that you can do for me.

Barry Benad: What? Just tell me.

CEO Ken: Now, I need you to go to China to the Cyberdroid Manufacturing plant.

Barry Benad: What do you want me to do or retrieve for you?

CEO Ken: I need you to find out the secrets about androids. Better yet, try to kidnap an android that is fully functional and ready for sale. I will examine her thoroughly, or take her apart if that is mandatory.

Barry Benad: I will try my best. I appreciate your wiring me the million dollars for the secret safe photos from the Cyberdroid Corporation. How did that work out for you? Did you ascertain anything about android creation?

CEO Ken: We did hit pay dirt. The only trouble was that for security reasons; the answers on android creation were written in some sort of personal code. Two of my best engineers could not decipher what meaning the personal code had about android creation.

Barry Benad: Oh, I am sorry to hear that. So close, and yet; so far…

CEO Ken: Yes, that keeps happening. The Cyberdroid Corporation is covering its tracks well.

Barry Benad: When do you want me to leave for China?

CEO Ken: I will think about it thoroughly and send you all the details. I have not had time to think about it yet. I just got back from the courthouse.

Barry Benad: Yes, I understand. Just call me back when you are ready. I hope you feel better. Just relax. Things will get better.

CEO Ken: Barry, I did not get this far in life just waiting for things to get better. I 'MAKE' things happen!

Barry Benad: Yes, ok. Call me soon. Have a nice day.

CEO Ken: Ok, Barry, bye.

79.) CEO Ken: (thinking out loud) Now, to think of a plan for Barry. Of course, he can add to the plan and refine it. It helps to write the plan down to ensure it will work.

   CEO Ken: (He wrote down all of the details. Read it. Then, re-read it.) There this will do. I see no loopholes. He emailed it to Barry who was living in France since the first attempt of theft at the Cyberdroid Corporation's secrets of Android Creation.

80.) The next day, CEO Ken took Chrissie out to lunch. They ate at one of Chrissie's favorite restaurants called Cem's.

Ken: Chrissie, you do look lovely today.

Chrissie: Thank you, Ken. I appreciate the compliment. (Chrissie smiles.)

Ken: What will you have today, Chrissie?

Chrissie: Well, the grilled chicken tender salad with romaine lettuce and cherry tomatoes looks delicious.

Ken: I think I will have the same.

Ken: (Ken summoned the waiter again by waving his hand at him. Sir, we are ready to order now. We both will have the grilled chicken tender salad with romaine lettuce and cherry tomatoes. Chrissie, what will you have for a drink?

Chrissie: I would like diet root beer, please.

Ken: Yes, that sounds delicious. I am so thirsty! Sir, two diet root beers, please. Could you bring them now? Thank you.

Waiter: Yes, sir.

Ken and Chrissie finished eating their salads, and drank the icy cool root beers.

Chrissie: Thanks for a delicious, nutritious lunch, Ken.

Ken: You are so welcome.

They walked back to the vehicle. They got into the cream-colored Leksus SUV to drive back to Collins' Corporation.

81.) In France at Barry's apartment:

Barry: (He opened the email from Ken Collins, CEO of Collins Corporation.)

Barry: (thinking) Hmm….this looks interesting. Well, it is a start. I have to think of the detailed actions. I will have to get Racio on a plane. He will be my accomplice again.

Barry: (on cell phone; international long distance call) ring…ring…Racio, this is Barry. I need you to get on flight 244 on Pan Am to China. Just go to the ticket counter. The ticket is pre-paid in your name.

Racio: Thanks, Barry. I will be on it. Don't worry.

Barry: Thanks, Racio. I will meet you at the Chinese Airport. Pack for a week's time of clothes. Wait in the food court at the airport. Take your cell phone so I can call you to find you.

Racio: Ok, Barry.

Barry: See you soon. Bye.

Racio: Bye.

The flights arrived safely at the Chinese Airport. Barry called Racio on his cell phone. They met again at the food court in the Chinese Airport. Barry walked up to Racio. Both Racio and Barry both stayed on the phone for directions as to where each other was located inside the Chinese Airport. It was lunchtime so they ate lunch together to discuss the plans in a far off corner of the airport where no one was sitting. They both had Mandarin Chicken with brown rice for lunch and iced tea to drink with NutraSweet sweetener and lemon wedges.

Racio: Ok, I understand the plan.

Barry: Thanks, Racio. We are a team. Sometimes, we are like Abbott and Costello. However, we get the job done.

Racio: Right, boss. Ha. Ha. Ha.

Barry: I have two hotel rooms adjacent to each other. One is for me. One is for you. Let us get some rest. We both have jet lag tiredness.

Racio: Ok, Barry. Good thinking.

The next day, Barry and Racio practiced the plan several times so that they would not mess up. Messing up could mean life in prison or death for both of them. There was no room for error. Not here. Not now! Two days elapsed since their arrival in China. Racio's attention span was running low.

Racio: Boss, I am tired now.

Barry: Yes, I know, Racio. However, we need to get this perfect. We do not want to be thrown in prison for life or executed.

Racio: Yes, you are correct, boss.

They had practiced for a week to get the heist perfect. Finally, Barry decided that they were ready.

They woke up at 2:00 a.m. Then, they dressed in their respective all black outfits with black socks and black tennis shoes. They ate sandwiches on whole wheat bread before leaving. Then, took care of their bodily functions before leaving, also.

They drove up to the Cyberdroid Manufacturing Plant in China. There was a mechanical roadblock with a guard sitting in a booth next to the mechanical roadblock. They pulled up to the security guard. Barry handed the security guard identification for both him and Racio.

Barry: We are exterminators. We have come to do the job at night. The smell will evaporate by morning.

Barry showed the security guard their identification for both him and Racio. The security let Barry and Racio have passes to enter the Cyberdroid Corporation. The mechanical roadblock lifted to allow them to enter the Cyberdroid manufacturing plant, which was located in China.

Barry: Barry rolled up the window and drove into the Cyberdroid manufacturing plant.

Racio: Well, that was easy.

Barry: Yes, just stick to the plan. It will be fine.

Racio: Yes, ok.

Barry: I will park the van here. Let us go in. I cannot wait to see the androids in their glass cylinders. I have read about them several times in popular magazines.

Racio: Yes, I will stick to the plan, but I am a little nervous.

Barry: It will be fine. Just stay with me, and stick to the plan.

Racio: Yes, ok, boss.

They both walked up to the magnificent entrance at the Cyberdroid Corporation. The Cyberdroid manufacturing plant looked like a high tech glass building. The glass was tinted a medium gray when viewing it from the outside. It was an extremely large building. There were signs above doorways. Barry had staked out this place once before with orders from Ken Collins. Barry knew the layout of the building and exactly where the finished prototypes of androids were being kept.

Barry and Racio walked down several long corridors. They encountered several security guards. All they had to do was to show their passes, identification, and their exterminator license cards. Barry had typed up all of the fake identification before they even practiced the present plan. All cards looked real.

Barry: Here it is, Racio. The reason we are here. The finished androids are here. They are held in tall glass cylinders with life support attached to them.

Racio: Ok, boss.

Barry and Racio found the finished product androids in tall glass cylinders with life support systems all around the androids. Barry looked at them all, and decided to kidnap Julie, a finished product android. Julie was manufactured beautifully. She had long thick blond hair with just a little bit of wave to her hair. Julie's eyes were a lovely medium blue/green color. She was built to perfection. Julie was 5'08" tall. Her body frame was medium. Her measurements were 36-24-36. Julie wears a 'B' cup bra. The Cyberdroid Corporation scientists downloaded a full Britannica Encyclopedia into her mechanical/human brain. Julie has maximum data storage available on any laptop or desktop computer. Julie is enabled to perform all mathematical formulas in her brain. She was designed to be a highly intelligent human being.

Barry and Racio put a tie of ribbon around her mouth so she could not speak. Then, they tied her wrist and feet. They put her in a large 4ft. x 4 ft. box. They made many air holes on all sides so she could breathe. They sat Julie upright in it. Barry instructed Racio to help him tape up the box with clear 2" wide mailing tape. They rolled her out of the building walking as quickly as possible. They did not want to look suspicious by running. They encountered one security guard.

Barry showed the security clearance passes, and professional exterminating licenses. The guard approved their departure.

Security Guard: Have a good night's sleep.

Barry: Thanks, I need it! (Barry smiled.)

Barry and Racio put Android Julie into the back of the white super utility vehicle (SUV). They closed the doors, and locked them. Then, Barry drove out calmly the same way he entered the Cyberdroid Corporation Android Manufacturing Facility. No one suspected anything. The next day scientist, Glenn Ganuchi, came to view and take care of the androids. He noticed android Julie was missing. Scientist Glenn Ganuchi immediately pulled down the burglar alarm siren. An extremely loud ear-piercing sound went throughout the building. He got on the intercom and issued a code orange of high danger. This is the most dangerous designated security idea. The burglar alarm contacted the local police station. Within minutes, there were 20 police vehicles in the Cyberdroid manufacturing parking lot. The Chinese police immediately swarmed out of the police cars like bees that had their nest attacked. They ran into the building to find scientist, Glenn Ganuchi. He told them that Android Julie was kidnapped. One police called the station to report putting up roadblocks all around the area within a 20-mile radius. There was an amber alert issued at all airports, train stations, and ships with Julie's photograph on it. Fortunately, Barry, Racio, and Julie boarded an airplane bound for the United States of America without being detected. Julie was loaded in the cargo area of the airplane. Barry made certain that the box had plenty air holes and that the box was well taped all around. It was a supersonic non-stop jet bound for the United States of America, which was capable of arriving in 1 hour.

Barry and Racio made it to the United States of America Atlanta, Georgia Airport safely. They waited to retrieve the cargo box in the designated cargo area. The 4 ft. x 4ft. cardboard box came down the cargo/suitcase mechanical runway. They retrieved their box and walked out of the Atlanta Airport as quickly as possible without appearing suspicious. Barry and Racio hailed a taxicab SUV. They placed the 4 ft. x 4 ft. box in the back of the van. Barry and Racio got into the second row of the taxicab SUV.

Barry: Please drive us to Dalton, Georgia bus station.

Middle Eastern SUV taxicab man, Abu Faruka: Yes, certainly.

After 2 hours, they finally arrived at the Dalton, Georgia bus station.

Barry: Thank you, sir. Here is the amount of the cab fare of $ 100 dollars plus the tip of $ 40.00.

Middle Eastern SUV taxicab man, Abu Faruka: Thank you, most gracious, sir. May you have a terrific day!

Barry: May you have a terrific day as well.

Barry called Ken on his cell phone. Ken gave Barry the address of a remote log cabin in the Northern Georgia mountains. Barry rented a dark green SUV from a car rental place across the street from the Dalton, Georgia bus station. They put the box inside of the back of the van. They opened the top to check to see if Android Julie was fine. She appeared fine. They took a chance because anything could have happened. The airport could have viewed the contents of the box for security reasons. The box could have fell open. It was an extremely dangerous traveling situation. However, luckily for Barry and Racio the airport was careless with the security precautions.

Barry drove to the log cabin located near Dalton, Georgia. Luckily, the SUV rental vehicle was equipped with a GPS (global positioning system) to ascertain the location of the remote log cabin easily. They arrived there in 10 minutes. Barry found the hidden key to the remote log cabin right where Ken told him to look. It was located in the hole of a tree trunk. Barry opened the door.

Barry: I am so relieved that everything turned out well.

Racio: Yes, I am relieved, also.

Barry and Racio carried the 4 ft. x 4 ft. wide box inside the remotely located log cabin.

They opened the box, and took a deep breath. They took Android Julie out of the box. They immediately plugged her into the wall socket to recharge, rejuvenate, and refresh herself. There were 2 – 4ft. high hard white plastic robots there in the log cabin. They maintained the log cabin and the grounds surrounding the log cabin to all boundary lines. Furthermore, the robots were equipped with strong lasers that can burn a person badly. The lasers shoot out of their chests near the shoulder part of the hard white 4 ft. plastic robots. Android Julie was informed that the two white 4 ft. hard plastic robots were equipped with strong lasers that can burn a person. Therefore, she should abide by the rules and not try to escape. They proved it to Android Julie when the two hard white plastic robots went outside and shot a tree that caught on fire. Barry had instructed Racio to hose the tree down to put the fire out.

82.) Ken immediately got into his SUV to travel to the remote log cabin. It was the weekend so he decided to spend it there at the remote log cabin. He arranged for the Collins Corporation scientists to conduct many tests on Android Julie without killing her. Ken, unlocked the door with his key that he keeps on his keychain. He walked inside the log cabin.

CEO Ken: Oh, she is absolutely perfect and beautiful!

Android Julie: You will never get away with this! The Cyberdroid Corporation scientists will find me. I have a GPS (global positioning system) inside of me.

CEO Ken: Have no fear! We only want to conduct tests on you without damaging or killing you. You will have no memory of this incident. You will have no recollection of this at all.

Android Julie: So, you say.

CEO Ken: I am a man of my word.

Barry: I can vouch for that. He paid me one million dollars as he said he would for a previous job.

Android Julie: That is comforting to know.

Barry, yes it is true.

CEO Ken: Just make yourself at home here in the log cabin. You are free to eat, watch television, play Nintendo games, watch DVDs, and do aerobic exercise. However, you must not use any kind of phone, cell phone, laptop, desktop computer, email, or any device that sends communications. You must not go outside, except when instructed to do so. Then, that will be on a chain for short walks around the premises. The 4 ft. hard white plastic robots will kill you with their strong lasers if you try to escape.

Android Julie: Yes, ok, I get the idea.

CEO Ken: Excellent! All of this will be over in a week's timeframe.

Android Julie: Thanks, I would like to start my life as the Cyberdroid Corporation intended for me to be. They find good homes, or set up a single life for androids with a secret identity that so one knows. An android's life must be kept secret because of the fact that human beings are jealous of androids. Humans claim that androids are taking all of the opportunities and jobs that rightfully belong to them.

CEO Ken: Yes, I know. Blake has thought of everything. He is an excellent creator of androids. Please feel free to eat anything. If you would like cooked food, please tell the robots. They will cook it for you. You are not allowed to cook or be around any potential for danger during this time period here. Do not worry! Just cooperate and all will be fine.

Android Julie: I wish I could believe you.

CEO Ken: You will. Trust me.

Android Julie: Ok, I will cooperate.

83.) At the Cyberdroid Corporation Manufacturing Plant in Beijing, China:

The burglar alarm siren was activated when Scientist Ganuchi noticed that fully functioning Android Julie was missing from the life support cylinder. Then, an automatic call went straight to CEO of the Cyberdroid Corporation's wireless international emergency phone. It can reach CEO Blake no matter where he is; even if he is taking a moon trip to the Earth's moon. In 2012, the elite and rich citizens can pay to take moon trips to outer space. CEO Blake has taken a few moon trips; however, when he received this emergency call, he had just returned from his weekend trip from Gatlinburg, Tennessee with his current fiancée, Jennifer. Blake answered the emergency call in his CEO office, which is located in Atlanta, Georgia, United States of America.

CEO Blake: Blake here.

Scientist Ganuchi: Sir, we have had a terrible tragedy!

CEO Blake: What? What is going on over there?

Scientist Ganuchi: Blake, a fully functioning, highly intelligent Android, Julie, has been taken. This is what we assume. We will question all of the security guards to see if anyone has entered the premise illegally or in disguise. It must have been in disguise because the burglar alarm system was not activated. I am assuming that the security guards gave a pass to bypass all other security guards. Blake, I will try to get a description of all characters that entered the premises last night after I left the building.

CEO Blake: Oh, how terrible! Android Julie is missing? Is that the one that I said was perfect, intelligent, and beautiful?

Scientist Ganuchi: Yes, the blond female.

CEO Blake: I wonder how they will use her. I hope it is not to disassemble her because she has a soul, too.

Scientist Ganuchi: Do not worry, Blake. I will be in touch again when I ascertain more about the situation.

CEO Blake: Thanks, I will be waiting for the call.

84.) At the Cyberdroid Manufacturing Plant in Beijing, China:

The police were swarming the Cyberdroid android manufacturing plant like bees on top of honey searching for clues. A top inspector of the police force, Inspector Cluenoo, questioned all of the security guards to take down their statement of any suspicious or new characters. When Inspector Cluenoo read the list, he noticed that the insect and pest control people were the only people let in the Cyberdroid Android Manufacturing Plant and one of the security guards gave him a pass to enter. Two other security guards saw him and another man walking down the hall. They asked to see his security clearance pass and let him through without any further problems.

Inspector Cluenoo spoke with Scientist Ganuchi in his Android laboratory. He told him that there were two insect and pest control guys who were given security clearance.

Scientist Ganuchi: That has to be the thieves! We have insect and pest control, but they come during daytime hours. We are supposed to have a lock down process at night. I want to have that security guard fired who gave those two thieves security clearance. CEO Blake trusted Scientist Ganuchi and let him have some control over the hiring and firing process

Inspector Cluenoo: Ok, I am glad I could get to the bottom of it. We will search on the security cameras to try to see their faces. However, from what the security guards told me; they were both fully covered in a white loose uniforms with a Honeybee masks on their faces.

Scientist Ganuchi phoned CEO Blake who was across the ocean at the Atlanta, Georgia, United States of America headquarters.

Scientist Ganuchi: Blake we have it narrowed down to this. Inspector Cluenoo told me all the details. It appears that the thieves were given security clearance by one of the security guards as I had suspected. They were dressed in beekeeper white uniforms with the masks. They used the cover that they were insect and pest control people.

CEO Blake: That should never have happened! We are supposed to be under high security lockdown at night always.

Scientist Ganuchi: Yes, I will have to fire the person responsible. He may have been paid to give them security clearance.

CEO Blake: Yes, that is what is sounds like to me. Thanks, Dr. Ganuchi for ascertaining what had happened. Still…we do not have any leads to try to find the thieves. Maybe, they will contact us for ransom money. It could be a kidnapping. We will have to wait and see now.

Scientist Ganuchi: Yes, I pray Android Julie is fine. I hope all will resolved with her safe return to the Cyberdroid Corporation.

CEO Blake: Yes, that is all we can do for now. Thanks, Dr. Ganuchi, I will keep in touch regularly.

Scientist Ganuchi: Ok, Blake, I will keep my fingers crossed and pray. Bye.

CEO Blake: Bye, Dr. Ganuchi.

85.) At the remote wilderness location in northern Georgia:

CEO Ken of Collins' Corporation hired three laboratory workers to test Android Julie. CEO Ken had her sedated to avoid her telling the laboratory workers that she was kidnapped. She was taken to a small l laboratory for x-rays of her entire body. Her blood was tested for many natural chemicals found naturally in the body. The scientists wrote a report about her and stated that she is a fully functioning android.

CEO Ken: Yes, I know she is a fully functioning android. However, what I want to know is what makes her have a soul. CEO Blake of the Cyberdroid Corporation said that he had a divine experience.

Kelly (laboratory scientist for Ken Collins): Yes, I know, but no one has ever been able to decode the soul secret. A judge granted CEO Blake and all androids full human rights in a Supreme Court Decision in the year 2000 because he said that they could fully think for themselves and display many human similarities.

CEO Ken: You mean I did this all for nothing?

Kelly (laboratory scientist for Ken Collins): Did what?

CEO Ken: Nothing. Just forget about it. Here is a check for the full amount of all of the x-rays and blood tests, $ 20,000 USD (United States Dollars).

Kelly (laboratory scientist for Ken Collins): Wow, Ken, that is $ 15, 000 USD over the costs.

CEO Ken: Yes, just do me a favor and keep all of this a secret.

Kelly (laboratory scientist for Ken Collins): Yes, thanks, Ken. We are bound by rules and laws not to discuss a patient's medical history with anyone else. Have a nice day! I must go now. I have another appointment shortly.

CEO Ken: Ok, thanks. You have a nice day, also.

CEO Ken picked up sedated Android Julie to transport her to his medium grey Laxus SUV (Super Utility Vehicle). CEO Ken laid her down on the back seat. He strapped her in with a seat belt. Then, he reclined the seat back to the position of lying down. CEO Ken drove android Julie back to the remote wilderness log cabin in northern Georgia, United States.

They arrived at the log cabin. CEO Ken carried sedated android Julie back into the log cabin and put her in a bed in the first bedroom. All of the windows had a special burglar alarm code to sound if any were opened. Therefore, android Julie could not escape out of the windows silently and secretly.

The next morning, android Julie awoke. She brushed her teeth; showered; and put on fresh clothes. She was excited because she knew if all tests were completed, she was free to be returned back to her creator, the Cyberdroid Corporation.

Android Julie: Sir, you said I would be returned unharmed after medical tests and x-rays had been completed.

CEO Ken: Yes, that is true. I am a man of my word.

Android Julie: Oh, how wonderful!

CEO Ken: We will have to think of a plan where we will not get into trouble. You do not know who we are; nor are you ever to reveal our identities if you ever ascertain our identities. Your RAM (Random Access Memory) will be erased from the last week of time.

Android Julie: Yes, sir. I promise. I am just glad to be returned unharmed and safe. However, I do not want to be taken again.

CEO Ken: I promise we will not get you specifically again if we ever do. You just keep a secret about our identities forever. You have done your service, and I appreciate that.

Android Julie: Oh, thank you, sir. I am so happy!

CEO Ken is a man of his word. He meant everything he just said to Android Julie. He made all the arrangements last night for her safe return. Android Julie was transported on a Collins' Corporate plane to Hawaii to be dropped off there. This way, that kind of cools the trail.

86.) At the Cyberdroid Corporation:

Ms. Jollyday answers the corporate phone.

Ms. Jollyday: Cyberdroid Corporation, may I assist you?

Android Julie: Yes, please may I speak to the CEO of the Cyberdroid Corporation?

Ms. Jollyday: No, it is not that simple. He has all of his calls screened first. He is a very busy man!

Android Julie: Yes, but this is important!

Ms. Jollyday: Your name, Miss, and the nature of this call?

Android Julie: My name is Julie. I need to come home.

Ms. Jollyday: Julie, what? Do you have a last name? What do you mean?

Android Julie: I do not have a last name.

Ms. Jollyday: Look, I have dealt with prank calls before, but this takes the cake.

Android Julie: Yes, it is true.

Ms. Jollyday: Ha ha ha! (Ms. Jollyday was laughing hysterically.)

Android Julie: No, it is true.

Ms. Jollyday: Now, look Miss, if you cannot cooperate by providing me with your last name; then you cannot make an appointment to speak with CEO Blake. We have security protocol here. I must hang up now.

Ms. Jollyday then hung up the corporate telephone at the Cyberdroid Corporation.

Android Julie called again.

Ms. Jollyday: Look, I can see your telephone number on the caller ID. I will have you arrested if you persist in tying up this telephone line.

Android Julie: (in a sobbing voice) Please, Miss, my home is the Cyberdroid Corporation. It is where I was born.

Ms. Jollyday: What did you say your name was again?

Android Julie: Julie. I do not have a last name as of yet.

Ms. Jollyday: Ok, just hold a minute.

Ms. Jollyday: Blake, I have this eccentric lady on the phone that says she wants to speak to you and that she was born at the Cyberdroid Corporation.

CEO Blake: Oh, my God! What is her name?

Ms. Jollyday: She just gives her first name as 'Julie'. She states that she has no last name yet.

CEO Blake: Put her through immediately and write down the phone number in case we get disconnected.

Ms. Jollyday: Ok, sir.

Ms. Jollyday to Android Julie: Ok, Miss, you will be put through now.

CEO Blake: Hello?

Android Julie: It is I, Julie.

CEO Blake: Julie, where are you? I will send for someone to come get you.

Android Julie: All I can remember is that I got off of a small charter plane. Now, I am in the Hawaii Airport.

CEO Blake: Oh, they must have erased part of your RAM (Random Access Memory). Do not worry you will get a full medical check-up to correct anything that needs fixing. Hawaii Airport?

Ok, just listen to me. Go up to the ticket counter and wait. I will have the local Hawaiian police there to get you temporarily. They will board you on the correct plane to Atlanta, Georgia, United States of America. I will pay a top police officer to escort you on the plane. Then, I will meet you at the Atlanta Airport. You must follow this security procedure for safety reasons.

Android Julie: Ok, I understand.

She did as she was instructed. Julie walked to the front counter. Julie told her story to the customer service person. The customer service person called security to inform them of what happened. A security guard came to wait at Julie's side until a police officer came to escort Julie back to Atlanta, Georgia, United States of America.

CEO Blake was waiting for Android Julie at the Atlanta, Georgia Airport in the United States of America.

87.) At Collins' Corporation:

CEO Ken: I just cannot believe my rotten luck! Twice we were successful in our attempts at thievery. Yet, we are still no closer to knowing the truth about the divine experience of Android Creation and what gives Androids souls.

Chrissie: Yes, I am sorry for you. However, I am glad that you did not harm the Android. There was no need for murder. Murder has a larger sentence to serve in prison or the electric chair. It is death, if caught.

CEO Ken: Chrissie, do not be ridiculous! You know that I am not a murderer. I may do some robbing to obtain corporate secrets; yet, I am no murderer!

Chrissie: Yes, just making sure and reminding you. Some people can get carried away in the heat of the moment.

CEO Ken: Yes, I know. Thanks for your concern. Anyway, still nothing. No secrets about android creation. All those medical tests and x-rays and still nothing to report.

Chrissie: Yes, I do feel sorry for you. It is tough being in second place in the information technology field.

CEO Ken: You do not know the half of it! I cannot tell you how much I loathe that!

Chrissie: Yes, I can imagine.

CEO Ken: No, really. There are no words to describe my hatred for CEO Blake Fencington and the Cyberdroid Corporation!

88.) At the Atlanta Airport, Georgia, United States of America:

    CEO Blake was waiting impatiently in the airport lobby at the Atlanta Airport, Atlanta, Georgia, and United States of America. He was waiting by Gate A, which is the gate for the Hawaii departures and returns. People started to disembark from the airplane. Approximately 20 people passed by in different nationalities and ages. Finally, a young, beautiful lady emerged through the crowd. As usual, Blake is stunned by her beauty. He remembers her from his last visit to the Cyberdroid Manufacturing Plant in Beijing, China.

CEO Blake went up to Android Julie with an extended hand.

CEO Blake: Ah, Julie, we meet again. I hope you had a pleasant trip back to the Atlanta Airport. Did all go as I had planned?

Android Julie: Julie touched CEO Blake's hand. Yes, all was fine. I am glad to be back with good people. CEO Blake walked with Android Julie by holding her hand down the long corridors of the Atlanta Airport. Finally, they got to the exit where Blake's Beep SUV was parked. CEO Blake opened the passenger front door. He let Android Julie get in the front. He buckled her with her seat belt.

Then, he got into the driver's side. Blake turned the Beep SUV into a small airplane just by pushing a button. The wheels of the Beep SUV pulled in. Then, some runway wheels and airplane wings came out of the bottom of the Beep SUV. Blake flew them back to the tall Cyberdroid Corporation, which has its own small landing pad for the corporate airplane on the roof of the building.

CEO Blake and Android Julie disembarked out of the special extremely fast corporate jet on the rooftop of the Cyberdroid Corporation. Then, they went into the Cyberdroid building and went into the elevator, which took them to the 20th floor of the Cyberdroid Corporation. Blake and Julie entered the Cyberdroid Corporation holding hands. Julie looked like she could be Blake's girlfriend. They looked like the perfect age and compatibility for each other.

They went into CEO Blake's corporate office. Julie sat across from CEO Blake in the client chair at his desk. Ms. Jollyday followed them into the CEO Blake's corporate office.

Ms. Jollyday: Well, is this the lovely Julie, I presume?

Julie: Yes, it is I.

Ms. Jollyday and Android Julie shook hands and smiled at each other warmly.

CEO Blake: Dr. Williams, please. (on the corporate international cell phone)

Secretary in doctor's office on phone: Dr. Williams' office, may I help you?

CEO Blake: Yes, this is Blake Fencington of the Cyberdroid Corporation. This is an emergency! I need to speak with Dr. Williams now.

Secretary in doctor's office on phone: Ok, just a moment. I will tell him you are on the line.

Dr. Williams: Hello, Blake, it is so nice to hear from you again.

CEO Blake: Dr. Williams, one of my androids has been kidnapped and released. I would like you to come and examine her fully at my corporate office. She has had too much excitement for the past week to go anywhere. I want blood tests of all kinds, and urine tests, also.

Dr. Williams: Ok, I will take my doctor's kit with vials and labels to store the specimens. I will be at your office after my last appointment today of 5 p.m. Therefore, I should arrive at your corporate office around 6 p.m. Is that fine?

CEO Blake: Yes, I understand you have to keep your appointments. That will be just fine. We will be waiting here for you. Just tell the doorman downstairs to ring me up here. I will give permission to let you up after hours.

Dr. Williams: Ok, Blake. It is always a pleasure doing business with you. We have been friends since high school. That is thirty years of friendship.

CEO Blake: Yes, I know. I appreciate our friendship. You have been a good friend to me through the years.

Dr. Williams: Thanks, and you, also. I will have to go finish up with my clients. See you later today.

CEO Blake: Julie, just make yourself comfortable here in my office. You are not to leave this office at all. You will be safe and secure here. I promise. My personal rest room is there. There are juices, teas, soft drinks, spring water, whole wheat bread, smoked lean turkey, and light mayonnaise in the small, glass refrigerator. There is a small television with some DVD movies, a laptop, and science and business magazines for your enjoyment over there in the corner of the room. The doctor will be here to collect specimens to make certain you are fine physically.

Android Julie: Ok, thanks. You have all the comforts of home here. I will try to relax. I appreciate your helping me, Blake.

CEO Blake: That is why I am here. You are in an excellent environment here. I will do everything in my power to try rectify everything for you.

Android Julie: Yes, I know. You are my creator.

CEO Blake: Yes, to a certain extent I am. However, the divine part that gives you a soul comes from the true Creator of the Universe. He has communicated to me what has to be done to make androids fully functional and self-thinking.

Android Julie: Yes, that is what I meant.

CEO Blake: Ok, you just do what I just told you. The doctor will be here at 6 p.m. today.

Android Julie: Yes, ok.

CEO Blake: I will just read some papers for the Cyberdroid Corporation here at my desk while we wait for Dr. Williams.

Android Julie browsed at all of CEO Blake's leisurely entertainment and decided to surf the Internet to view clothes online and read the ABC news online.

89.) Time elapsed. Dr. Williams arrived promptly at 6 p.m. eastern standard time zone.

Dr. Williams: Could you please tell CEO Blake that I am here now? (He told the doorman downstairs.)

Doorman Henry: Dr. Williams is here to see you, Blake.

CEO Blake: Yes, please let him come up. Give him a permission pass.

Doorman Henry: Ok, sir. You have a good day!

Doorman Henry: Ok, Dr. Williams, take this permission pass in case you get stopped by a security guard on your way up to the 20th floor of CEO Blake's office.

Dr. Williams: Ok, thanks.

The doorman unlocked the glass door to let Dr. Williams into the Cyberdroid Corporation.

Dr. Williams went to the elevators and took the elevator up to the 20th floor. He walked out of the elevator towards the glass that had the Cyberdroid name on it. Dr. Williams walked inside. He was greeted by the Cyberdroid secretary, Ms. Jollyday.

Ms. Jollyday: Dr. Williams, Blake is waiting in his office for you. You may enter.

Ms. Jollyday: (on speakerphone) Blake, Dr. Williams is here.

CEO Blake: Yes, show him in.

CEO Blake: Dr. Williams, thanks for coming. Julie, Dr. Williams is here. He will give you a checkup now.

Dr. Williams: Julie, you can stay in that chair. I will just collect some blood samples for some tests. Then, you can take this cup into the bathroom with you and fill it with some of your urine.

Android Julie: Ok, I will cooperate.

Dr. Williams collected the 4 – one ounce blood samples. Then, he gave Julie the urine sample cup to take with her into the bathroom alone.

Android Julie went into the bathroom to urinate and caught some urine from the urine stream before it went into the toilet. She wiped herself. Then, she washed her hands with anti-bacterial soap. She glanced at herself briefly in the mirror to see if she looked alright, and emerged into the CEO corporate office again. Julie gave the plastic, clear urine sample cup with the lid on it to Dr. Williams.

Dr. Williams: Thank you, Julie. I will do some tests and let you know how you are doing as soon as possible. I will drop these off at the hospital laboratory on my way home from work. Blake, it is always good to see you. It is late. I have appointments early tomorrow morning. I need to get my rest. Call me to do a round of golf, tennis, or racket ball at the gym.

CEO Blake: Yes, Dr. Williams. Thank you for coming. I will certainly take you up on your offer.

They shook hands, and smiled at each other. Dr. Williams left the CEO corporate office.

CEO Blake: Julie, it is time to go home now. You can sleep in the guest bedroom alone tonight at my home.

Android Julie: Thank you, Blake. I appreciate that.

   90.) The whole Cyberdroid office was dark and quiet. There were only hall lights on for the security guards to tour the building at different times for security. Ms. Jollyday left for the day. No one was there except the security guards in the security guard lounge area. Blake took Android Julie home with him. He was a complete gentleman. Android Julie stayed in a guest bedroom. She felt safe and comfortable there.

   91.) The next morning Android Julie came into the kitchen for breakfast with CEO Blake. He drank some juice and took a generic Sentrum multivitamin. Blake gave Android Julie the same thing. Android Julie took her vitamin as well.

CEO Blake: Julie, as you know I am CEO of the Cyberdroid Corporation. I lead an extremely busy life. That is why I cannot be your guardian here. You need to be flown back to Beijing, China where the people at the android manufacturing plant can see to it that you are purchased to be with a respectable family, respectable man, or in an individual setting with a job. That is not my specialty. I just lead the corporation. Therefore, today you shall be flown back to Beijing, China. I only want the best for you, Julie. I hope you can understand that.

Android Julie: Yes, I want to be put somewhere in society in a peaceful, respectable setting.

CEO Blake: I promise you, Julie, you will be put in a wonderful, respectable, peaceful place. You can come to visit me in the future anytime. Just make an appointment with me by calling first.

Android Julie: What I truly want is to be independent with a good job to live near you here in Atlanta, Georgia.

CEO Blake: Well, if that is what you truly desire, you shall have your wish. That is the least I can do since you have had unfortunate circumstances with the kidnapping. Thank God, you are safe and unharmed.

Android Julie: Yes, that is what I desire.

CEO Blake: Ok, your wish will come true. I will speak to Ms. Jollyday today so she can speak with the human resources department here at the Cyberdroid Corporation. We will find an opening here for you to work. Then, Ms. Jollyday will find a suitable apartment for you with security. I will obtain your social security card. You will have to pass the driving requirement to receive a driver's license. I will get all the mandatory identification requirements. Now remember, Julie, your identity as an android is to remain top secret under all conditions. Many people are jealous of androids. Some say that androids are taking humans' jobs, etc. Furthermore, you do not want to be kidnapped again. Your safety is in secrecy of your true identity as an android.

Android Julie: Yes, I understand.

CEO Blake: I will make us each a 98% lean sliced turkey sandwich on whole wheat bread. How does that sound?

Android Julie: That will be fine. I appreciate all you do for me.

CEO Blake: You are welcome. Now, let us eat breakfast. I have to go to the office after eating. You can come with me. Ms. Jollyday will take care of all that I have just described to you. You will spend the day with her and another worker. She will get someone to take you to take the driver's license test today. Within a couple of weeks, you should have your own apartment, job, and start on a single life here in Atlanta, Georgia.

Android Julie: Ok, that would be wonderful!

CEO Blake: Ok, let us eat. Then, I will have to finish dressing for the office. Then, I will drive us there in my SUV Beep vehicle.

92.) They left promptly at 7:30 a.m. eastern standard time zone. CEO Blake drove them in his SUV Beep vehicle.

CEO Blake: Julie, fasten your seatbelt, please.

Android Julie: Ok.

They arrived at the Cyberdroid Corporation at 8:00 a.m. eastern standard time zone.

CEO Blake: Ms. Jollyday take care of all identification needs for Julie. Hire her at the Cyberdroid Corporation in an office position opening. Keep Julie's android identity a secret. Get someone to help you out with taking her to take the driver's license test.

Ms. Jollyday: Yes, I will do all that needs to be done.

CEO Blake: Thanks, Ms. Jollyday. I can always count on you.

Ms. Jollyday: It is always a pleasure working for you, Blake. By the way, Blake, did you happen to see the news last night?

CEO Blake: No, why?

Ms. Jollyday: Well, there were protests in front of the Beijing, China android manufacturing plant. Many people are protesting that they do not want androids in society. Some said on the news camera when interviewed that the androids are taking the human beings' jobs, and other opportunities that were meant to be truly for humans only.

CEO Blake: Did scientist Ganuchi call here?

Ms. Jollyday: Yes, he was frantic. He said he had to call the police because the crowd outside was too large.

CEO Blake: That was the correct thing to do. I will call scientist Ganuchi about this. Oh, as you know, Ms. Jollyday, that humans are jealous of androids. It is imperative that androids keep their android identity a secret to avoid being ostracized by society. Many people have grown accustomed to the android idea. Regardless, the Supreme Court of the United States ruled that androids have all the rights of human beings since they passed the test to have a soul, think independently, and look just like humans.

Ms. Jollyday: Yes, I can understand all of that. In the future, many more people will think of androids in a better light. I hope and pray anyway.

CEO Blake: Yes, I do, too. Now, I will call scientist Ganuchi at the android manufacturing plant in Beijing, China. I am not worried. The Cyberdroid has all of the top security, motion detectors, cameras, and security guards. Security has been heightened since Android Julie's kidnapping.

93.)

CEO Blake: (on phone) Mr. Ganuchi, this is Blake. I am just calling to tell you that you did the correct action by calling the police. Large rowdy crowds can do much damage. You do not have to worry, Mr. Ganuchi, security was heightened ever since Android's Julie's kidnapping. We have found her and she is doing fine. She will reside here in Atlanta, Georgia. Her android identity will remain a secret. Ms. Jollyday is helping me to get her settled in an apartment, a job here at the Cyberdroid Corporation, and all of her identification cards. I discussed it with android Julie and it is her wish to reside here in Atlanta, Georgia as a single lady. That is the least I can do for her since her ordeal of the kidnapping.

Scientist Ganuchi: Yes, I fully understand. I wish her the best of luck with her new surroundings and future. Give her my regards.

CEO Blake: Yes, I will tell her that you wish her good luck with her future. Keep up the excellent work, Mr. Ganuchi. Keep in touch regularly. Chat again soon.

Scientist Ganuchi: Ok, bye.

94.)

CEO Blake: (on phone) Jennifer, this is Blake. How have you been doing? I have had some emergencies in which to deal. However, now all is under control.

Jennifer: Oh, that is good. I was beginning to wonder what happened to you.

CEO Blake: Jennifer, you know I care deeply for you. We are engaged now.

Jennifer: Yes, I know. I love you, too.

CEO Blake: Would you like to go out to dinner tonight?

Jennifer: Yes, I would love that.

CEO Blake: How about dinner at 6 p.m. after work?

Jennifer: Yes, that is fine.

CEO Blake: Ok, see you later.

Jennifer: Ok, bye.

95.)

CEO Blake: (on intercom) Ms. Jollyday, would you come into my office, please?

Ms. Jollyday: Yes, certainly.

   Ms. Jollyday walked into CEO Blake's office.

CEO Blake: Ms. Jollyday, I have not spent any time with my fiancée, Jennifer, ever since all of this drama with the kidnapping. Everything is under control now. Would it be all right if Android Julie stayed with you, for a while? It would not be for long. She will be in her own apartment soon as you already know.

Ms. Jollyday: Yes, it is fine. I can understand that you need to spend time with your fiancée, Jennifer. My husband is easygoing. He will not mind. She will have her own room with a lock on the door. Furthermore, you know I have a burglar alarm.

CEO Blake: Yes, Ms. Jollyday. I know you have ABT wireless burglar alarm security as I do. I know android Julie will be fine there. Oops, we need to call her just, 'Julie'. She is entering into society now. We do not want to reveal her android identity.

Ms. Jollyday: Yes, I will just call her 'Julie' from now on.

CEO Blake: Thanks, Ms. Jollyday. I need to relax my nerves. It has been so hectic for the past few days.

Ms. Jollyday: Ok, relax and enjoy yourself. You deserve to relax after all that has happened.

CEO Blake: Thanks, Ms. Jollyday. That will be all for now.

96.)

References

Nordin, P., Wolff, K., Tallhamn, M., & Kihlmann, M. (2000). Priscilla Android Photo. Chalmers

University. Retrieved from Http://www.androidworld.com.

www.ingramcontent.com/pod-product-compliance
Lightning Source LLC
Chambersburg PA
CBHW041950220326
41599CB00004BA/173